Sound for Moving Pictures

Sound for Moving Pictures presents a new and original sound design theory called the Four Sound Areas framework, offering a conceptual template for constructing, deconstructing and communicating all types of motion picture soundtracks; and a way for academics and practitioners to better understand and utilize the deeper, emotive capabilities available to all filmmakers through the thoughtful use of sound design.

The Four Sound Areas framework presents a novel approach to sound design that enables the reader to more fully appreciate audience emotions and audience engagement, and provides a flexible, practical model that will allow professionals to more easily create and communicate soundtracks with greater emotional significance and meaning.

Of obvious benefit to sound specialists, as well as motion picture professionals such as film producers, directors and picture editors, *Sound for Moving Pictures* also provides valuable insight for others interested in the subject; such as those involved with teaching soundtrack analysis, or those researching the wider topics of film studies and screen writing.

Neil Hillman, PhD, is a highly experienced sound designer, production sound mixer and scholar who contributes to both multi-million budgeted feature films and crowd-funded indie movies. His PhD investigated emotion within sound design and considered enhanced audience engagement through the thoughtful use of the soundtrack.

Sound Design

The Sound Design series takes a comprehensive and multidisciplinary view of the field of sound design across linear, interactive and embedded media and design contexts. Today's sound designers might work in film and video, installation and performance, auditory displays and interface design, electroacoustic composition and software applications, and beyond. These forms and practices continuously cross-pollinate and produce an ever-changing array of technologies and techniques for audiences and users, which the series aims to represent and foster.

Series Editor: Michael Filimowicz

Foundations in Sound Design for Interactive Media
A Multidisciplinary Approach
Edited by Michael Filimowicz

Foundations in Sound Design for Embedded Media
A Multidisciplinary Approach
Edited by Michael Filimowicz

Sound and Image
Aesthetics and Practices
Edited by Andrew Knight-Hill

Sound Inventions
Selected Articles from Experimental Musical Instruments
Edited by Bart Hopkin and Sudhu Tewari

Doing Research in Sound Design
Edited by Michael Filimowicz

Sound for Moving Pictures
The Four Sound Areas
Neil Hillman

For more information about this series, please visit: www.routledge.com/Sound-Design/book-series/SDS

Sound for Moving Pictures

The Four Sound Areas

Neil Hillman

Routledge
Taylor & Francis Group

LONDON AND NEW YORK

First published 2021
by Routledge
2 Park Square, Milton Park, Abingdon, Oxon OX14 4RN

and by Routledge
52 Vanderbilt Avenue, New York, NY 10017

Routledge is an imprint of the Taylor & Francis Group, an informa business

© 2021 Neil Hillman

British Library Cataloguing-in-Publication Data
A catalogue record for this book is available from the British Library

Library of Congress Cataloging-in-Publication Data
Names: Hillman, Neil, author.
Title: Sound for moving pictures: the four sound areas / Neil Hillman.
Description: New York: Routledge, 2021. | Series: Sound design | Includes bibliographical references and index. |
Identifiers: LCCN 2020046825 (print) | LCCN 2020046826 (ebook)
Subjects: LCSH: Motion picture music–Instruction and study. | Film soundtracks–Production and direction. | Sound design.
Classification: LCC MT64.M65 H55 2021 (print) | LCC MT64.M65 (ebook) | DDC 781.5/42–dc23
LC record available at https://lccn.loc.gov/2020046825
LC ebook record available at https://lccn.loc.gov/2020046826

ISBN: 978-0-367-51779-3 (hbk)
ISBN: 978-0-367-51778-6 (pbk)
ISBN: 978-1-003-05518-1 (ebk)

Typeset in Times New Roman
by Deanta Global Publishing Services, Chennai, India

Access the companion website for support material at: www.soundformovingpictures.com

To my Mother and Father, who love me but remain puzzled by what I actually do to earn money; and for Heather, Toby, Sam and Caitlin: my four areas of emotional focus.

Contents

Figures

Tables

Foreword

Sound for moving pictures

In 1802, to the initial bafflement of the scientific community, a lecture on *The Modification of Clouds* was given at the Askesian Society, in Plough Court just off Lombard Street in the City of London. The lecturer, Luke Howard, was a 31-year-old chemist of no previous distinction, but for this presentation he had branched out to meteorology—the science of weather—which had been his obsession since childhood. The unlikely goal of the lecture was the scientific classification of clouds, something that very few until that evening had considered possible, or if possible, of little interest. By the end of Howard's lecture, the possibility had been confirmed, and interest forcefully ignited.

What transfixed the audience was Howard's proposition that clouds have many *shapes* but only a few basic *forms*: that fundamentally there were only four *genera* of clouds: Stratus, Cumulus, Cirrus, and Nimbus, distinguished by their distance from the surface of the Earth, and whether they were actively precipitating or not. The great variety in the shapes and activity of clouds—cirrocumulus, stratocumulus, cumulonimbus, cumulocirrostratus, etc.—comes from the unlimited combinations of those four fundamental *genera*.

Several months later, Howard's lecture—expanded and illustrated—appeared in Andrew Tilloch's *Philosophical Magazine*, the best-known science journal in Europe at the time, and today, more than two hundred years later, Howard's four categories still underpin every atmospheric study and all weather reports on the evening news.

Like clouds, the swirling sounds of nature and the thunderous man-made noises that surround us have been our constant but mostly disregarded companions since time immemorial, and like clouds, sounds have a shifting infinity of impermanent shapes that seem to defy categorization.

But Edison's 1877 invention of the means of recording and reproducing sound; and then fifty years later the perfection of our ability to synchronize a sound recording with moving pictures; and quickly thereafter to layer multiple sounds on different tracks; and then to shift the synchronization between those sounds and the image; all these inventions sharpened our appreciation of sound, and gave us the ability—the urgent need, in fact—to figure all this out.

Figure 0.1 Cumulostratus with Cirri above, from Howard's Modification of Clouds

Béla Balázs, at the dawn of film sound in 1930, put it succinctly and poetically:

> The vocation of the sound film is to redeem us from the chaos of shapeless noise—to reveal for us our acoustic environment—the acoustic landscape in which we live, the speech of things and the intimate whisperings of nature; all that has speech beyond human speech, and speaks to us with the vast conversational powers of life.

In *Sound for Moving Pictures*, Neil Hillman makes a convincing case for four *forms* of sound—*Narrative, Abstract, Temporal*, and *Spatial*—underlying the infinite *shapes* that sounds can assume. These categories are second cousins once removed of Howard's *Stratus, Cumulus, Cirrus*, and *Nimbus*.

For administrative purposes, we mixers and sound designers normally distinguish between Dialogue, Music, and Sound Effects (DME), but these rigid distinctions are there as much for legal and technical purposes as anything else, and they can limit our thinking about what a broader consideration of sound categories can accomplish: they deafen us to *all that has speech beyond human speech.*

Twenty years ago, I proposed a more flexible theoretical spectrum than DME, stretching from what I called *Encoded* to *Embodied* sound. Dialogue was in a broad cluster at the encoded end, music spread across the embodied end, and sound effects scattered in between as sound-centaurs: part linguistic and part musical.

Hillman's *Narrative* and *Abstract* categories roughly correspond to my use of Encoded and Embodied, but he extends that linear distinction into the third dimension with two additional categories: *Temporal* (how rhythmic a sound is) and *Spatial* (how reverberant it is, and/or in what quadrant of the theatre it is placed). Like Luke Howard's hybrid clouds (cirrostratus, stratocumulus, nimbostratus, etc.) this added dimensionality allows for multiple entanglements of the four basic categories: temporal–narrative, spatial–abstract, narrative–temporal–spatial and so on, depending on the dominant characteristic: a sort of hierarchical nephology of sound.

Dr Hillman's book is an extension of the PhD thesis he wrote a number of years ago for a Creative Practice programme. This thesis had in turn profited from his forty years of experience in the practice and aesthetics of recording, editing, and mixing for film—including running his own post-production company, which has branches in the UK and Australia.

Hillman writes that his interest in theory emerged out of his real-world experience in weaving complex soundtracks. But why would an experienced, established practitioner like Hillman venture into the sometimes remote academic world of theory?

In 2020, the creative use of sound for films is barely older than a single human lifetime—93 years—and it has been a fast-developing but still-imperfectly understood art compared to the traditional ones—painting, drama, dance, music, and so on—whose origin, tens of thousands of years ago, is lost in prehistory. As Balázs noted: *Cinema is the only art whose birthday is known to us.*

Moreover, it happens that there is a deep neurological imbalance in the human brain: 30 per cent of our neurones are dedicated to processing visual information, but sound occupies barely a tenth of that amount. This disparity is reflected in the dozens of vivid words we have in English and other languages for different ways of looking—*glimpse, stare, peep, gaze, ogle, etc.*—and the paucity of words for listening: we have only two—*hear* and *listen*—and neither of those carries any emotional colour.

So we are relatively inexperienced in this art of sound for film, and culturally impoverished in our tools to analyse it; therefore, any increase in our understanding should be welcome. But precisely how can theoretical categories like Hillman's be practically useful to the often visceral, schedule-dominated world of sound production?

They are above all a help in *creation*—by encouraging the dissolution of sclerotic conventional assumptions about the relationship between sound and image and revealing new possibilities. My own experience has been that audiences quickly grasp unusual and metaphoric juxtapositions—in fact they seem to hunger for them—and we sound designers can always go further in stretching categories than we may first think. A rigid dialogue–music–effects approach is frequently a straight-jacket which does not empower the creative process, however much it may serve the administrative needs of the system.

Hillman points out two other benefits: theoretical categories can be *analytical* tools for academics and historians in their studies of film soundtracks; and categories can facilitate the sound department's *communication* with directors and producers, providing the terms to clarify and articulate the needs of sound, for the good of the film.

I have dwelt on Hillman's theories in this foreword, but the majority of the text you are about to read is also thankfully rich with concrete and detailed practical analyses of film soundtracks: some that were important to Hillman's creative development—Bergman's *Winter Light*, Coppola's *Apocalypse Now*, and Lars von Trier's *Dogville*—and then a number of films from Hillman's own work, where he has inside information about creative and technical struggles.

In the closing pages of the book, Hillman cites Steven Spielberg: *The eye sees better when the sound is great*; and George Lucas: *Sound is half the experience, where you get the most bang for your buck.*

These observations implicitly acknowledge the interdependence of sound and image: how the audience's focus of attention can be artfully guided by sound in both subtle and overt ways.

At the opposite point of the director's compass from Spielberg and Lucas is Andrei Tarkovsky. His feelings, which Hillman quotes, put more emphasis on the value of sound in and for itself:

> The sounds of the world reproduced naturalistically in cinema would be a cacophony [...] I have a feeling that there must be other ways of working with sound, ways which would allow one to be more accurate, more true to the inner world which we try to reproduce on screen; not just the author's inner world, but what lies within the world itself, what is essential to it and does not depend on us.

There is an echo here of Balázs's observation, which I have already cited:

> to reveal for us the speech of things and the intimate whisperings of nature; all that has speech beyond human speech, and speaks to us with the vast conversational powers of life [...] The vocation of the sound film is to redeem us from the chaos of shapeless noise.

As different as the films of Lucas/Spielberg and Tarkovsky are, there is a commonality underlying their ideas about sound for motion pictures, but the breadth of that commonality is vast, and Hillman's book reveals the scope of what the soundtrack for motion pictures, pushed to its aesthetic and technical heights, can achieve, if we are prepared to do the climbing.

Hillman's theories, and the examples that he gives, are significant rungs on the ladder to those heights.

Walter Murch

'Wait a minute, wait a minute, I tell ya, you ain't heard nothin', yet!'

Al Jolson - *The Jazz Singer* (1927)

Preface

This book is at its heart a love-letter to cinema, and in particular cinema sound, thinly disguised as an investigation into the relationship between the creative process of designing and mixing moving picture soundtracks, and an exploration of the emotions elicited in listening-viewers by doing just that.

It is something that I've spent a fascinating, challenging, stimulating and exhausting working lifetime getting to grips with (literally and figuratively) as a location sound recordist, a dialogue and ADR editor, a supervising sound editor, a sound designer, a re-recording mixer and an Outside Broadcast sound supervisor.

Now approaching the fourth decade of my career as a sound practitioner, the first three fit into roughly equal parcels of activity.

I consider it extremely fortunate that I was able to change track early-on in my working life, switching lanes from the shop floor of the machine tool industry to the studio floor of the film and television industry; because enjoyable as it was in its own way, I had increasingly come to realize that whilst a career in the new technology of Computer Numerical Controlled (CNC) milling machines might have fed me, it would not nourish me; and so within a short time of completing a comprehensive four year apprenticeship in mechanical and electrical engineering, and graduating with a foundation degree in electronics, I became an employee in the sound department of a major UK broadcaster, Central Independent Television plc; initially as a trainee at their Elstree Studios near London, and then as a sound technician in Central's Birmingham city-centre studios.

Central TV inherited all the privileges of the previous and long-time Midlands ITV franchise-holder, ATV – a broadcasting, feature film and music publishing giant – and it therefore sat as an upstart newcomer at the programme-planning top-table of the 'Big 5' ITV companies; who in those days (along with Granada, Yorkshire, Thames and London Weekend Television) vied constantly for output on the independent commercial network. Central was a hugely productive and extremely profitable company that eventually (but with hindsight, inevitably) succumbed in 1994 to a combination of shareholder greed and the financial temptations of a free-market economy, espoused by Mrs Thatcher's Big-Bang Britain.

This is a time looked back on by some of my peers as a kind of 'golden age' of broadcasting and programme making, and certainly it was as much fun as it was hard work; but viewed now, little of the content we produced at the time really stands up against the overall quality that is taken for granted today. (If only that same sense of fun and *esprit de corps* remained on film sets, or in television studios today … .)

I am, however, eternally grateful for the training I received from Central Television, which covered every aspect of sound operations, both on film and video; and this also included audio equipment maintenance, which happily meant I was able to utilize some acquired knowledge from my academic work in electronics, as well as the time I had spent as an apprentice electrical and mechanical engineer.

Joining Central in the early 1980s as a trainee, with some experience of presenting on university radio and a few spots at an Independent Local Radio station under my belt, I mistakenly thought that this, and my qualifications, would somehow stand me in good stead. Instead, I was aware that any mention of academia was frowned upon in the blue-collar atmosphere of the television studios, which the technicians I now worked with called 'the real world'. Yet here I was in a workplace that was as far from the real world as I could possibly imagine; certainly, compared to the Machine Tool factory environment that I had been working in for the previous six years.

I'd replied to an advertisement for sound trainees and attended an interview; however, through some kind of administrative error, the job offer that eventually came through the post to me from Central was as a trainee studio camera operator, rather than a trainee sound operator; but not wishing to push my luck ahead of the scheduled medical and induction day, I sent my acceptance, turned up, got checked for colour blindness and then promptly presented myself in the Sound Department. It worked – nobody questioned me being there. I was in.

At this time, location drama and documentary recording work were undertaken by the Film Department; of which the film cameramen were the undisputed glamour-boys, adored by the likes of the female make-up artists (and some of their male colleagues, too). It seemed to be an accepted fact that all cameramen were attractive; and if film cameramen were the equivalent of *Top Gun* fighter pilots, then the studio cameramen were almost as fascinating as the *Memphis Belle*'s bomber pilots. But those of us who worked in sound – either location or studio – well, we were the tail-gunners … Which reminds me of a cruel – but nonetheless funny – cartoon strip I once saw in a trade magazine. A morose guy is alone in a bar, late at night, bemoaning his situation to the barman:

Drinker: 'Women don't find me attractive at all … OK, I admit I collect baseball cards and I know I have a terrible haircut but come on – I once won an Oscar!'

Barman: 'You have an Oscar? Oh man, chicks love that kind of thing – what did you win your Oscar for?'

Drinker: 'Dialogue editing.'

Barman: 'Talk to me about those baseball cards…'

So, I quickly learned that sound was not an attractive area to be known for; but I also knew, absolutely, that I had found my true home and my life's calling.

When I joined the industry, nothing had really changed about the technology in years, aside from the introduction of colour television pictures; and with the poor resolution of hand-held video cameras at that time, the TV station that I worked at had as many film cutting rooms equipped with Steenbeck 16mm flatbed film editing tables, as it did 2-inch videotape editing suites.

News reports were shot on cheaply developed reversal film, and often shot mute, with reporters voicing-over items live in the studio that 'grams operators' would add audio atmospheres to, from quarter-inch tape cartridges as they saw the pictures play for the first time, live on air. The imaginative use of a selection of 'carts' containing white noise, city skyline and pedestrian footsteps managed to cover most stories in those days (clearly, this was an early introduction for me into sound design … But nobody called it that, back then). Negative film was more expensive to develop and so it was reserved for drama and feature work and continued to be so until such time as portable video cameras were of sufficient picture-quality to make more widespread appearances on location.

At this time, location filming for sound recordists involved utilizing mono, quarter-inch Nagra audio tape machines that recorded audio with a 50Hz pilot tone, for faithful playback-speed purposes; and then back at base, the tapes would be transferred to magnetic, sprocketed film, for assistant film editors to lay-up for pre-mixes, and projectionists to load on to multiple sound-followers, allowing film dubbing mixers to create pre-mixes, taking their cues from carefully coloured, paper dubbing charts, showing the footage number at which transitions occurred. These charts could be works of art that either helped or hindered a mix, depending on the attention to detail that the assistant film editor had applied to their sheets. It is still a source of personal pride that my first audio post-production role was as what I thought of as a 'proper' dubbing mixer, i.e. I was working on *film*.

Within Central's television studio complex, banks of 2-inch helical-scan analogue Video Tape (VT) machines recorded the vision-mixed studio output, with sound from the studios fed to the VT machines via assignable tie-lines, selected by a mechanism based on the antiquated but efficient Strowger electro-mechanical telephone exchange multi-selector; as fascinating a piece of engineering that the nineteenth century had to offer …

The story goes that Almon B. Strowger was an undertaker in Kansas City, USA, and there was a competing undertaker whose wife was an operator at the local telephone exchange. Whenever a caller asked to be put through to the Strowger Funeral Home, calls were deliberately put through instead to his competitor. This obviously frustrated Strowger greatly and in 1888 he patented a system for doing away with the human part of the routing equation, thus ensuring that any calls meant for him, were routed directly to him (Michael, 1996). Incredibly, a century later, the audio paths in some television studios were still being assigned in this manner.

At that pre-digital time, whether working on film, or video, both technologies had a fundamental post-production requirement: the editing of pictures and sound demanded a reliable method for synchronization between the two – and the pictures and sound that we now digitally de-couple and re-lock at will with our Non-Linear Editing systems and Digital Audio Workstations had rather more rudimentary locking mechanisms then: film through the well-established medium of sprocketed magnetic stock, video through what were state-of-the-art electronic Adam Smith 2600 synchronizers, that could slave two Studer 24 track audio tape machines to a single master VT machine thanks to an SMPTE timecode reference signal.

I remember well what felt to me as the white-knuckle peak of analogue audio and video synchronization, when too close to transmission to complete a layback, a master Ampex VT machine a whole floor away was 'locked' to the two Ampex 24-track audio tape machines in the audio post-production suite; ensuring that synchronous sound and pictures went out live to the nation, from two totally separate areas of the studio building.

That programme was *Spitting Image* (1984), and equally cutting-edge in its time. It was a production of such audio complexity (e.g. track count) and topicality, that it meant that two dubbing mixers, plus an assistant, (sometimes me) required a week of exceptionally long days to deliver each 23-minute episode. Today, with the digital technology we now have at our fingertips, and given the way Production Executives have sharpened their pencils on production budgets, I would suggest that one sound editor and one dubbing mixer would be expected to turn such a show around within three days.

I left Central – reluctantly, but I hoped for sensible commercial reasons – in early 1990, before the unprecedented and industry-wide redundancies that followed. I left freely, without a redundancy pay-out, but in doing so I bought myself time to make headway at the outbreak of a revolution in the UK television industry; one that was inspired, encouraged and led by Mrs Thatcher's dim view of the medium. More precisely, what the Prime Minister seemed to take most objection to was the working terms and conditions that protected television staff at that time.

The first shots had been fired on September 21st, 1987, when the Premier hosted a seminar at Downing Street and shocked broadcasting industry leaders by declaring that: 'Television is the last bastion of restrictive practices' (Brown, 2013).

Therefore, within the industry itself, it was increasingly apparent that things were about to change; but it took until 1991, with a by then deposed Thatcher watching from the side-lines, for the ITV franchise process to signal a sea-change in the working conditions and remuneration for British film and television technicians.

What followed the re-allocation of the regional broadcasting licences was a wholesale shedding of expensive, in-house staff; and an explosion in the growth of independent production companies that were highly profitable at the cost of anxious and exploited freelance workers engaged by these new companies (Silver, 2005).

So began my second decade in the industry; which was spent as a freelance sound technician, working predominantly on location as a recordist, and mainly for video; but there were still a good number of film assignments along the way: for 35mm movies, a rarefied and heady environment for any young professional, it was only 2nd unit work – on features such as *On Dangerous Ground* (1996) for instance; but 16mm jobs were still more than plentiful in the 1990s. I worked as the Recordist on cinema commercials and also on documentaries shot on film, such as Bill Bryson's travel series *Notes From a Small Island* (1999), and on numerous corporate productions (frequently for the Halifax Building Society, who at that time made good use of a more than £1 million annual filming budget to shoot high-quality training films, *on film*) and of course, the new kid on the block at this time was the pop-promo. These were copious, too; with, memorably for me, several shoots to promote tracks from UB40's 1993 album *Promises and Lies* in the mix.

As a Location Sound Recordist, my television clients were for the most part the major UK broadcasters; with BBC Birmingham's *Top Gear* (1978)[1] and the BBC's Science and Features Department in London – the home of long-running series such as *Horizon* (1964), *Tomorrow's World* (1965) and other one-off science-based documentaries – particularly helpful in enabling me to see the far-corners of the world, all at British television licence-payers expense. I travelled overseas extensively during this period, so much so that I was issued with two passports to ensure that visas could be issued whilst I was abroad; and also to ensure that stamps and visas from sensitive countries could be kept tactfully away from equally sensitive immigration officer's eyes.

Another regular BBC client during this time was BBC Bristol's *Antiques Roadshow* (1979), for whom I introduced MS stereo location recording to the programme; and I happily remained responsible for the production's location sound for five consecutive series.

My third decade commenced with me returning to a studio environment, as Head of Sound for a picture post-production facility that had decided to increase its offering to include audio post-production. I was an operational Head of Department, with responsibility for the departmental staff as well as the quality of its audio output and broadcast compliance.

It was here that I first became known no longer simply as a 'Supervising Sound Editor' or a 'Dubbing Mixer' (UK industry terms), but also as a 'Sound Designer' and a 'Re-recording Mixer' (US terms) through carrying out both of those roles on an American co-production; a 26-part, half-hour children's animation series, somewhat unattractively called *Butt-Ugly Martians* (2001). It was actually great fun and tremendous experience. To get the series under way, I got to work on the first two episodes alongside the hugely experienced Supervising Sound Editor, Rick Hinson (*The X Files* (1993), *Buffy the Vampire Slayer* (1997), *The Sopranos* (1999)). Rick was not only a Primetime Emmy winner for his work on *The X Files* but also great company and generous with his advice. I learned fast and I learned a lot from him.

The schedule was punishing (the time difference between the UK and the West Coast of America regularly extended our working day, sometimes by a further eight hours) and we were working to the exacting requirements of the brilliant US Producer Bill Schultz (ex-*The Simpsons* (1989) and *King of the Hill* (1997)). 'BUMS' – as the series became affectionately known amongst those of us working on the production – was described as 'the most watched TV animation series ever' (I'm fairly sure *The Simpsons* would go on to put that BUMS publicity claim in the shade), but our work on *Butt-Ugly Martians* was recognized by the British Royal Television Society, who rewarded our efforts with an 'On Screen Excellence' award.

Following completion of the busy *Butt-Ugly Martians* series, I spent a year working closely with the RAF *Red Arrows* on a Discovery Channel documentary called *Beyond The Horizon* (2000); a series that boasted 'unprecedented access to the world-famous display team and their support staff' (the company I was working for were nothing if not up-beat with their hyperbole) and I recorded the sound of 'The Reds' on location in the UK and in their pre-season preparations in Cyprus; and then edited and mixed in post-production, the behind-the-scenes material we had recorded with them.

The experience of that year became the catalyst for me to understand that only by running my own audio post-production facility could I ever ensure the commitment to quality that I sought; one that would emulate the no-compromise ethos of those magnificent men in their flying machines (the Red Arrows' motto is *éclat* – 'brilliance') and this is how my audio post-production company, *The Audio Suite*, came to be born.

Remarkably, The Audio Suite would go on to have more credits listed on *IMDb* than that 'Big 5' ITV company I started my television career with; and although now reined-back and by choice more considered in the kind of work we do, the studio's output in our first decade as a small, operator-owned audio post-production facility was nothing short of prolific: for broadcast programming alone, we averaged one hour of national network broadcast television, every week, over each of our first ten years of trading.

That was by any definition an incredible achievement for us as a small, Birmingham-based, independent audio post-production facility; especially given that Birmingham had diminished from being a major television production centre throughout the 1960s, 1970s and 1980s, to virtually nothing by the mid-1990s. This disappointing state of affairs – driven by deliberately divisive BBC and ITV operating policies – resulted in an almost overnight creative exodus, and left the spaces those screen companies had occupied quickly looking down-at-heel: an empty, shabby-chic that can be seen in Steven Spielberg's *Ready Player One* (2018), a movie that created much local excitement when it came to Birmingham to film.

Part of my aspiration for The Audio Suite was to create a 'studio sound' that would be as distinctive and recognizable as the major American film studios had been to me as a child; when I was able to discern audible differences between them, without really knowing why. It was only later that I discovered that this was largely due to the repeated use of their 'stock' sound effects, such as galloping horses, gunshots and bullet ricochets, that either the Warner Bros or MGM sound departments took from their own effects libraries. (The familiar sounds of Cowboy films and Wild West soundtracks that we would watch as kids at Saturday morning film clubs like the 'ABC Minors';[2] in my case, held at the Sutton Coldfield *Empress* cinema.)

My hometown's elegant mock-Tudor High Street (which in the past was graced by a *real* Tudor, Henry VIII, whose extensive hunting grounds remain and are now known as Sutton Park) was subsumed in the early 1970s by the multiplying metropolis of Birmingham. It was re-developed immediately after its acquisition, and quickly rendered anonymous; but before this post-Brutalist architecture depressed the town like a permanent wet weekend in November, I remember it being book-ended by cinemas: at the north end was *The Empress* (quickly reduced to rubble by developers keen to provide a hardcore base for a bland Sainsbury's supermarket) whilst to the south, at the top of the hill, was the *Odeon* (still operating as a cinema, but now a multi-screen part of the *Empire* chain and disappointingly dishevelled inside. If I ever win a substantial sum of money, I intend to immediately buy it and return it to its former glory as a single screen theatre).

As a young child in the late 1960s I remember seeing matinee performances at the Empress, such as *Half a Sixpence* with my mother, and *Where Eagles Dare* with both of my parents; and then significantly, as a teenager whose 'newspaper round' money was used to fund regular visits to 'the pictures', the art of kung fu found me when I saw *Enter the Dragon* at the Odeon.[3]

I could detect differences between American animation studios by their soundtracks, too – the style of comedic effects used in Hannah Barbera cartoons such as *Huckleberry Hound*, *The Flintstones* and *Tom and Jerry*, were markedly different to those in United Artists' *The Pink Panther*, which were different again to UPA's *Mr. Magoo*; but all of them subconsciously watermarked my early listening.

Much later, my private quest for an Audio Suite signature-sound felt somewhat pretentious, given that the commissions to create bespoke movie soundtracks were only lightly sprinkled among the factual television entertainment that dominated our studio schedule (but thankfully, paid the bills); and I was incredibly grateful that in the early days of starting the studio I was entrusted to edit, mix and deliver 40 hours of network programming with a travel series called *Home from Home*. It got my studio under way and the momentum it created continued unabated for a dozen more years.

We grew quickly from being a studio into *studios*, and along with sound designing for British micro-budget and passion-project films at one end of the spectrum – and providing ADR services for major Hollywood features at the other – we became best known for our high integrity, tight turnaround television soundtracks: over 120 hours of a motoring magazine show called *Fifth Gear*, and more than 220 hours of a consumer technology programme called *The Gadget Show* were delivered to UK broadcasters, and then re-purposed for international syndication through the *Discovery* channel.

The only consistent creative space available to me became the sound of our narration – the presenter's commentary – and I attempted to make these recordings distinctive from any other studio by virtue of their 'effortless intelligibility', their tone, their presence, and their pacing and spacing within the allocated gaps on the timeline – a radio station approach to television. Our standard set-up for narration became (and remains mine for narration work) a large diaphragm condenser, cardioid microphone set 8 to 12 inches away from, and just above being in line with, the presenter's mouth; processed by a modern, outboard microphone pre-amplifier that modelled a vintage compressor, set at a 3:1 ratio, and driven relatively gently by the input signal, to provide 3–4 dB's of gain reduction.

Whilst commercials and late-night radio love the use of the 'warm, dark' proximity effect of a cardioid microphone worked close to a voice artist's mouth, I wanted a more natural sound; one that made the viewer feel that the narrator was sitting next to them in their living room, as they watched the programme.

It was satisfying being recognized by our peers, too, with several nominations (including one for the coveted UK Conch Award for 'Facility of The Year') as well as actual award wins; including two World Medals in Sound Design from The New York Festival, a Gold from the LA Film Festival and various Royal Television Society awards, including one for 'Best Production Craft Skills'.

These trophies have a place in the studios that I call 'the ego shelf', with absolutely no disrespect meant towards the awards themselves. Instead, it is a reminder to keep things in perspective and not to get too carried away either with a sense of self-importance, or to take for granted the capricious nature of the industry and the projects that manage to find you. Although it is unlikely I ever will, I would like to think that I would feel exactly the same about winning an Oscar. (Two degrees of separation and all that: I was fortunate to hold one of *Lord of The Rings* Re-recording Mixer Mike Hedges' several statuettes once, on a visit to his beautiful art-deco mixing theatre in Wellington, New Zealand; housed within Peter Jackson's impressive Park Road Post facility.)

I grew up in a monochrome, monophonic 1970s Britain, where coal miners strikes and the imposition of a three-day working week made us children move from our sprawling comprehensive schools to being taught in church halls; when power cuts meant quiet evenings spent in candlelight and widespread industrial unrest, an oil crisis and winters of workers' discontent set the backdrop for the mood of the nation. Television was the staple entertainment and, poor though it was in retrospect, its popularity had grown increasingly at the expense of cinema.

As Dominic Sandbrook notes:

> By 1970, only 2 percent of the population went to the cinema once a week, compared with a third in the late 1940's. During the next 10 years, audiences continued to collapse: by 1980, total admissions had fallen by half, dipping below 100 million for the first time since the early years of moving pictures. (Sandbrook, 2011, p.45)

He goes on to describe how the cinemas themselves were singularly uninspiring, run-down and sad, with their formerly elegant façades of art deco and worn-through velvet upholstery patiently awaiting their retirement fate as Bingo Halls. At this point, it seemed likely that cinema going would die out entirely in Britain.

But it was television that would provide the unlikely life-line for both the British Film industry and the cinemas themselves in the early 1970s.

As Andrew Higson observed, although 1971's biggest box-office hit was Disney's *The Aristocats*, in second place, from the genre of TV sit-com spin-off, was a vastly different offering: a cinematic version of Thames Television's *On The Buses*. A contemporary working-class pastiche, it was by no means unique; that year *Up Pompeii* and *Dad's Army* successfully made the transition from small to big screen, with several more titles following throughout the decade: *Steptoe and Son*, *Mutiny on the Buses*, *Please Sir!*, *Love Thy Neighbour*, *Porridge*, *The Likely Lads* and *Rising Damp*. All these films were united on two counts: they brought cinema audiences back; and they were pale imitations of the originals (Higson, 1994).

I know this last point to be true because I was there: at the Odeon cinema, in Sutton Coldfield. It was 1973, I was in my first teenage year, and to the flickering light of a matinee screening of *Holiday on the Buses*, a blushing girl from my class at school, Heather Reinman, and an awkward me, made an impromptu pairing in support of our respective best friends, who were going out on their first date. Afterwards, my date and I agreed that the film was pretty poor (and suspected that our friends saw little or nothing of the presentation from lights-down to lights-up), and other than our faltering conversation that mainly served as a critique on contemporary British cinema, it's fair to say that due to our shyness and the situation, that lovely girl and I hardly got to know each other any better. For us at least, love was not in the air that afternoon.

But like cinema itself our time was yet to come; and that same girl and I went on to live totally separate lives, with many air miles, hemispheres and new citizenships of distance between us. Until against the odds, after 25 years apart, we re-met and fell hopelessly in love; and 37 years after we had sat through the adventures of Jack, Stan and Blakey on the buses, we married. The setting we chose for the ceremony was in our hometown of Sutton Coldfield; and as it happens, the wedding venue was not too far from that handsome Odeon cinema.

That then, is the back-story.

However, the origin of this text lies not only in my nearly four decades as a curious and questioning, humble foot soldier–practitioner (although it is heavily influenced by this). It is also wrapped up in a quest to seek, formulate and catalogue a new way of considering sound design; reflecting and formalizing an approach that I'd unconsciously picked up and developed over the years, and which led me on a journey to research a PhD by Practice part time, whilst continuing to practice as a Sound Designer and Re-recording Mixer, full time. And this book is a vehicle to share where my heart and mind come to rest, with regard to creating soundtracks for moving pictures; whether they are for viewing on a hand-held or a laptop screen, in the intimacy of a home living room, or for experiencing in the immersive, social environment of a multi-channel movie theatre.

When we synchronize pictures and sound, we use the term that sound is 'married' to the pictures; but like all healthy marriages, the two individuals are interesting entities and possess depth and complexity in their own right. And it is in the union of the two that their true potential is fulfilled, enabling both to fully blossom. I unashamedly use this analogy for motion picture production.

What I increasingly wanted to understand as a jobbing Sound Designer and Re-recording Mixer, was how the sound for moving pictures, be it a soundtrack for film or television, works on our emotions; as well as on our conscious and sub-conscious; and I wanted to know if it could be shown that a skilled audio practitioner – a Sound Designer or Re-recording Mixer – could somehow shape an audience's emotional reactions to the pictures they are seeing.

Some of the assertions I read in Norman Holland's 2009 book *Literature and the Brain* really intrigued me, particularly when he said:

> The brain's tricks become clearer at the movies. The cute blond starlet, looking for her missing friend, opens a creaking door. She walks down a dark hall. And we're thinking, 'Don't go there! Don't go there!' And then the maniac in the hockey mask lunges out from a dark corner, brandishing a chain saw. You jump and I jump and all the people around us jump. Yet you and I and all of us know deep down that the blond and the maniac are just light flickering on the screen. We still jump – why?
>
> (Holland, 2009, p.3)

This is as obvious an example of the hand-in-glove way that sound and vision work together to elicit an emotional response from the audience that you are ever likely to see. The creaking door accentuates our nerves, already aroused by the darkness of the hall, which of course offers us no clue as to what the starlet will find. The pace of the cutting and the rhythm of the audio also need to tease the audience along; the time between on-screen events needs to be extended to heighten the expectation that something is about to happen. *We see the darkened hall* – pause to allow our imagination to conjure our own vision of what lies within the darkness – *cue the sound of the creaking door* – pause again to give space to our rising instinct to run – *Crash, Bang, Wallop!* – the audience has leapt out of their seats by the image that just jumped onto the screen, accompanied by prominent Foley and a signature sound: in this case the stylized revving motor of a chainsaw …

That example is a tried and tested method of leading an audience through sound and vision to a predictable, and fairly obvious, conclusion. We want the audience to jump out of their skins at a given point in the film, and they do; and this works very well for the Horror genre, or for suspense thrillers. But surely, I thought, there are many other emotions that are available to be touched. After all, as the audience enters into the theatre, does it not also enter into an agreement with the filmmaker to be, amongst other things, moved, inspired, uplifted and entertained?

So when I read the results of work which showed that listeners can *infer* a speaker's emotion from auditory cues, independent of the meaning of the words uttered, it seemed entirely possible to me that moving picture soundtracks could also be designed in such a way as to intentionally influence the emotional state and attitude of its listening-viewers, independent of the story or visuals that they are watching.

Therefore, in the research work I did for my PhD (no need for a plot-spoiler alert: the resulting thesis was snappily entitled 'A new sound mixing framework for enhanced emotive sound design within contemporary moving picture audio production and post-production'), I set out to investigate how certain aspects of audience emotions could be enhanced through specific ways of mix-balancing the soundtrack of a moving picture production; primarily with the intention to intensify the viewing, and listening, experience for the audience.

By re-visiting my PhD research notes, considering the work of other sound theorists and practitioners, describing my own novel sound design framework and then critically reflecting on my personal creative practice whilst using that framework, in this book I have set out to demonstrate how different ways of balancing a moving picture soundtrack can influence an audience's emotional response; and that the Four Sound Areas framework that I propose is a new and useful way to look at things when either *constructing* a soundtrack from the point of view of its emotional outcome for moving-picture creatives, *deconstructing* a soundtrack for academic or professional analysis or in *communicating* a desired or intended emotional outcome to fellow moving picture professionals: Producers, Directors, picture editors or indeed, other sound practitioners.

As a teaching aid, I have also briefly appended each chapter with one or two suggested discussion points that might naturally arise from that chapter's content, to assist those who might consider using this work as either the basis for delivering a multi-session module, or for individual lecture planning; and the topics contained within each chapter can in turn either find merit as the basis for an assignment that encourages further investigation on the topic from students, or act as a useful primer for the practical exercises that are also suggested: encouraging all flavours of film students to experiment further – on their own, or in a group exercise.

As this book draws on my PhD research work, it is entirely appropriate to acknowledge the huge debt I owe to my PhD Supervisor, Dr Sandra Pauletto, for her invaluable support throughout my entire research project, resulting in the successful completion of my PhD at the University of York's department of Theatre, Film and Television (TFTV) (now the department of Theatre, Film, Television and Interactive Media).

I consider myself to be extremely fortunate to have worked with such a supportive Supervisor as Sandra, whose interest in not only the subject of my thesis but also my wider work as a practitioner, managed to somehow instil a confidence in me to complete that body of work whilst I continued my challenging day (and night) job in a busy, demanding, unpredictable and exhilarating freelance sound environment. Now teaching in Sweden at the KTH Royal Institute of Technology, I am happy that our friendship continues.

In this vein, I must also extend my grateful thanks to Dr Jez Wells from the University of York's Music department and Dr Dave Payling from the Film and Journalism division of Staffordshire University's School of Computing and Digital Tech, for their help in achieving a kind of clarity that had proven to be somewhat elusive at times.

For those readers who are students continuing to work after a Bachelor's or Master's degree for their own doctorate, I hope that this book may also serve as an inspiration for you to keep going. I urge you to keep going and 'keep the faith'; it is likely to stretch you further than you ever imagined.

A PhD is by its very nature intrinsically a solo pursuit – a passage into the unknown; and my part-time, self-funded PhD was not unlike (in my imagination at least), a single-handed circumnavigation of the globe in a yacht. But as with any such endeavour (notwithstanding the personal commitment to finish what was started, and the necessity of riding out the daunting and dispiriting Southern Ocean of pre-*viva voce* self-doubt), an undertaking of this sort would be utterly impossible without the help of others, outside of the academy. I must therefore also thank – profusely – the special people who helped keep me afloat when, metaphorically, the wind was out of my sails and psychologically, I was bailing water. (As anyone who has stayed the course will tell you, completing a PhD is not just character forming, it is character revealing.)

First and foremost, I must offer my often inadequately expressed gratitude to my wife. The unstinting support I have received from Heather over the years, coupled with her perennial patience, interest and topical suggestions for my consideration, is nothing less than saintly. Already attuned to losing her husband for periods of time calibrated in 'Dubbing Mixer Units' (DMU – an hour to a Dubbing Mixer immersed in their work is a loose, nominal measurement of time that ranges variously between one and 24 hours), my decision to undertake a PhD by Practice that would for over nine years neatly fill the gaps in our already rare spare time enjoyed together, was greeted with her characteristic reaction: encouragement. Heather, you are my rock.

I feel blessed that my parents were able to see their son receive his doctorate; and my Dad and Mum, Aneurin and Beatie, have over the years remained a constant source of grounding (in a good way). As are my three, incredibly precious children: Toby, Sam, and Caitlin (and what fun it was to still be engaged with my university course whilst you three entered, graduated and moved on with your own continuing professional development. Who knew that having a Dad who could produce an NUS Student Discount Card to a waiter in a hip restaurant could be simultaneously 'hilarious', 'cool' and 'sad'?)

Last, but not least, I owe a special thanks to my stepdaughter Nicola in Australia (who herself is one of three siblings – along with Nicola, Heather is a devoted mother to Nicola's elder sister Jemma and their younger brother, James). Thank you Nic for providing the distance, space and peace to write-up so much of my PhD thesis in your Emerald, and then Brisbane, homes; the former an exciting and exotic experience of living in the baking heat of the Tropic of Capricorn, augmented by many joyful and treasured memories of the happy distractions provided by our two gorgeous grandchildren, Hudson and Bethany, coming to find their Poppa after his sitting for too long on his own, hunched over a laptop computer.

I recognize that I am a lucky man indeed to be blessed by such a family; one that is unavoidably extended yet surrounded by loving kindness. The recent additions of another grandson, James' son Theo, and another granddaughter, Jemma's daughter Mary, has simply made the distance that separates us from our Australian contingent all the more acute at times.

During my PhD research period, and then through the time further researching for the writing of this book, I have also enjoyed the support of colleagues who have come to recognize that once again the topic of conversation has been steered towards subjects vaguely allied to my field of interest, study and research; and many a mixing stage (whilst we wait for a pesky Pro Tools-rig to re-boot), a film location catering bus (whilst we wait for the rain to stop) or an Outside Broadcast vehicle's control room (whilst we wait for 4k Ultra HD pictures to be synchro-nized) have staged debates on the widest of audio matters; and all totally worthy of their own 'TED Talk'.

My underpinning motivation for undertaking a PhD by Practice on an unor-thodox approach to sound design, and then writing a book about its results, is undoubtedly rooted in my life-long love of cinema; which later developed into a fascination (bordering on obsession) for the sound that accompanies the pictures.

I have been privileged to meet (and with several, discussed at length my thoughts on moving picture sound design) such luminaries as Walter Murch, Gary Rydstrom, Randy Thom, Eddy Joseph, Michael Hedges, Robin O'Donaghue and David Macmillan; and together, they form a collective of talent, wisdom and experience that has amassed a total of 17 Oscar wins, 35 Oscar nominations and a further 40 BAFTA wins and nominations (at the time of writing) for feature film sound recording, design and mixing.

Their generosity in both listening, and then corresponding to continue our dis-cussions further, reinforces them to me as cinema 'greats' in every sense of the word; and I offer my deepest personal and professional respect – and grateful thanks – for their input, influence and above all else, their inspiration. I hope that I have gone at least some way in their eyes to fulfilling this subject's undoubted potential.

So now, let us begin this journey into deep sound design; and in the spirit of *Star Trek*'s Captain James T. Kirk, let our mission be 'to boldly go where no man has gone before': in search of *The Four Sound Areas*.

N.H.
Moseley,
June 2020

Notes

1 I am grateful to the *Top Gear* presenter Jeremy Clarkson not only for the great fun and friendship of filming many items on location with him over the years, but also for creating the opportunity to write my first mainstream press article – the 'guest column'

in the May 1997 edition of *Top Gear* magazine. Filming together to cover the launch of the new Audi A4 car, over several days on the beautiful island of Sardinia, each night after work he and I would retire to our rooms to write our respective other commitments: his for the *Sunday Times*, mine for the trade-press, *Studio Sound* magazine. Over pre-dinner drinks in the bar one night, we compared our writing fees; and when he had finished laughing at his favourite topic – someone else's misfortune – Clarkson took pity and offered to get me a 'better-paid gig'. He was true to his word; and that *Top Gear* column still remains my best paid return on writing.

2 The ABC Minors was a Saturday morning film club for children, run by the ABC Cinemas chain. Established in 1927 by John Maxwell, by 1945 ABC had over 400 picture-houses, second only in the UK to the Odeon/Gaumont chain of theatres. Like British cinema in general it declined during the late 1950s and early 1960s, but remarkably, its lineage still exists – unrecognizable, but alive in spirit – through myriad and convoluted sales, mergers and acquisitions from the 1970s to the present day, having being the property of familiar cinema names such as the Associated British Picture Corporation, Warner Bros, Severn Arts, EMI, Cannon Cinemas, Pathé, MGM Cinemas, Virgin Cinemas, UGC and Cineworld.

3 Bruce Lee and *Enter the Dragon* (1973) cemented my life-long interest in practising martial arts, the seeds of which were sown earlier, by watching *Kung Fu* (1972) on television. I still imagine myself as I did at 13 years of age: stepping in as the body-double for seven-times World Professional Karate Champion, Chuck Norris.

1 Introduction

1.1 Introduction

This book is presented as a comprehensive body of work for those engaged in moving picture production; particularly professionals whose practice incorporates an involvement with sound, *and* the topic of sound design; either in a creative sense, or as a method of critical analysis of the soundtrack for wider film studies purposes.

But it is also arranged in such a way as to aid academics and students interested in the topic of moving picture soundtracks, either as a reference work for sound design research purposes, or as a teaching aid. The book may be used as a standalone work, or ideally, in conjunction with the media that is available on the companion website found at www.soundformovingpictures.com.

The book provides a three-tier consideration of sound design in contemporary moving picture production and post-production, using an original, practical and theoretical methodology called the Four Sound Areas framework.

Firstly, the Four Sound Areas framework is shown as a way for steering and intensifying the intended emotional experiences of listening-viewers, by describing a method with which a Sound Designer[1] might consider *constructing* a soundtrack; and a Re-recording Mixer[2] might subsequently build on this desired emotional impact, by emphasizing specific elements of the soundtrack during the final mixing stage.

Secondly, the Four Sound Areas framework is described as a means of *deconstructing* the soundtrack, to allow the academic, student and critic to carry out a more detailed analysis; the purpose of which being to arrive at a clearer understanding of artistic audio motives, and the narrative significance of sound, in any moving picture production.

Finally, the Four Sound Areas framework is presented as a tool for creative professionals to *communicate* emotional intent for a soundtrack; to more easily enable, e.g. a film Director to express their desired emotional intent to a Sound Designer, a Re-recording Mixer or a picture editor.

Central to this whole process is my proposal that all moving picture soundtracks, such as those for filmed entertainment or broadcast television, are created from an audio 'compound' made up of four distinct elements, termed the *Narrative, Abstract, Temporal* and *Spatial* sound areas; and that these areas form a useful framework for both the creation and the consideration of emotional sound design.

The *Narrative* sound area is concerned with sound that carries direct communication and meaning. Dialogue and commentary are the most important examples of this area, which also includes symbolic and signalling sounds such as ringtones, sirens and other sounds; and, music with a clearly defined meaning.

The *Abstract* sound area is concerned with sounds that have a less codified and clear meaning. Atmospheres, backgrounds, room tones, sound effects and music are examples of these.

The *Temporal* sound area is concerned with the evolution in time of the sound design. Its characteristics are rhythm, pace and punctuation. This area can include music, sound effects and voice.

The *Spatial* sound area is concerned with the positioning of sounds within a three-dimensional soundfield *and* the space placed around the presented sound.

I have used the specific term 'listening-viewer' throughout this work when referring to the intended audience of a soundtrack, to differentiate from the more general and widespread media use of the terms 'viewer' and 'listener'. The word 'balancing' refers to the 'level balance', or 'mix balance' between the discrete soundtrack elements that constitute the compound of a moving picture soundtrack; the manipulation of which is under the complete control of the Re-recording Mixer. Greater detail regarding the Four Sound Areas is contained in Chapter 4 of this book.

A short animated film, *Jack – Safe@Last* (2010), a short feature film *The Craftsman* (2012), a full-length feature film *Here and Now* (2014) and a recording of a live sports Outside Broadcast event *Commonwealth Games Boxing* (2014) are used to demonstrate the Four Sound Areas in use in my own commercial work, and excerpts of these may be found on the book's companion website, www.sou ndformovingpictures.com.

1.2 The Sound Designer and the concept of sound design

Gary Rydstrom, a renowned Sound Designer and Re-recording Mixer at Skywalker Sound, considers his role to be:

> someone who the Director turns to as being in charge of the soundtrack. (Loewinger, 1998)

With this in mind, in creating, constructing and organizing a soundtrack, it is the job of the Sound Designer and the Re-recording Mixer often – but not always – the same person[3] (especially if mixing 'in-the-box')[4] to understand, create and manipulate the inter-relationship of the proposed Four Sound Areas of this study: the Sound Designer presents a *selection of sounds* considered capable of creating emotional impact and narrative consequence, whilst the Re-recording Mixer determines *the relative balance* of these sounds (i.e. the emphasis given to each of the Four Sound Areas within the mix at any one time), to elicit, emphasize or steer the emotional response of the listening-viewer at any given time during the programme.

It is also worth defining at the outset, those pockets of 'ring-fenced' autonomous creative practice that exist for the Sound Designer and Re-recording Mixer; as well as the similarities and differences in the responsibilities and objectives of these two different roles. The first similarity and overriding purpose of both jobs is to serve the film's artistic and technical potential; but how they go about achieving this is somewhat different.

The Sound Designer initially identifies the 'extra' sounds that are required to augment or even replace the supplied production sound (supplied by the Production Mixer, usually via the picture editing department, which normally consists of only the dialogue recorded on location) and then presents to the Re-recording Mixer, all and every sound element that they have interpreted to be of significance to the plot, the storyline or the emotional intent of a scene. This audio arc encompasses production sound, dialogue editing, ADR recording, sound effects, Foley, atmospheres and music.

Tarkovsky notes:

> As soon as the sounds of the visible world, reflected by the screen, are removed from it, or that world is filled, for the sake of the image, with extraneous sounds that don't exist literally, or if the real sounds are distorted so that they no longer correspond with the image—then the film acquires a resonance. (Tarkovsky, 1987, p.162)

Typically, the Sound Designer's output comprises dialogue, atmospheres, sound effects, Foley and score; and they may also synthesize and manipulate sounds to create new, unique sound effects. In a word, the Sound Designer is committed to providing aural *content*. The Sound Designer's position in the sound team is highly creative and autonomous, but can also carry significant logistical responsibility for managing the audio post-production team, its budget and its processes if the fuller duties of a Supervising Sound Editor are also undertaken.

The Re-recording Mixer must sort through the myriad supplied audio, which by this stage consists of Dialogue and ADR (Automated Dialogue Replacement), Foley, Sound Effects, Atmospheres and Music. They need to understand the intention and motive for each element of the presented soundtrack, determining, enhancing and controlling the tonal quality, sonic fidelity and volume level of the individual audio clips, whilst providing objectivity as to what should be included or discarded to best serve the emotional needs of the movie. This rationalization is an essential, and highly creative process. Tarkovsky comments again:

> The sounds of the world reproduced naturalistically in cinema are impossible to imagine: there would be a cacophony. Everything that appeared on the screen would have to be heard on the soundtrack, and the result would

amount to sound not being treated at all in the film. If there is no selection, then the film is tantamount to silent since it has no sound expression of its own. (Ibid.)

A Re-recording Mixer is required to work quickly, accurately and to a high standard. In short, the Re-recording Mixer is committed to providing the *context* for the Sound Designer's *content*. A Re-recording Mixer requires a great deal of manual dexterity, as well as craft skills, technical knowledge and emotional sensitivity. The role offers a great deal of creative freedom, but it also carries total responsibility for the nature and quality of the final soundtrack.

With regard to the creative freedom bestowed onto both the Sound Designer and the Re-recording Mixer, Deutsch observes:

in practice it's common that a considerable amount of sound post-production (effects creation, dialogue replacement, track-laying, processing, pre-mixing, etc.) is done away from the Director's oversight, and is a reflection of the distinctive craft and creativity of the sound department. (Deutsch, 2018)

The full extent of this creative freedom away from the Director's eyes (and ears) is worth stating; and is made manifest by the amount of work that has to be carried out to get the sound stems into a ready-to-be-mixed (or pre-mixed) state. Whilst a modern Digital Audio Workstation (DAW) bestows great flexibility for the manipulation of sound clips, there are still routinely up to 14 decisions a sound editor must make for each and every audio clip on the timeline, e.g.:

1. Open head of clip
2. Open tail of clip
3. Shorten head of clip
4. Shorten tail of clip
5. Fade in clip
6. Fade out clip
7. Increase or decrease clip-based gain
8. Cut within the clip (then repeat all the above)
9. Crossfade between clips
10. Pan
11. Apply clip-based EQ
12. Apply clip-based Reverb
13. Remove location room Reverb
14. De-noise.

The audio 'clip-count' is high in modern movie production and the 'clip tally' makes interesting reading: a 20-minute, 5.1 'reel' can easily accommodate 3,000 audio clips. With 14 decisions required per clip, each 20-minute reel can contain in the region of 42,000 decisions. An average feature film of 100 minutes has

5 reels; therefore 3,000 clips, each requiring 14 decisions, across 5 reels, can result in around 210,000 creative decisions being made – in the absence of the Director – even before the mix.[5]

As the soundtrack is dissected in this book, it considers and explores all the key components of sound design (i.e. voice, sound effects, atmospheres and music) and several different recording, processing and reproduction techniques; with a view to providing for the reader a wider understanding of the role a Sound Designer and Re-recording Mixer can play in evoking emotional responses in a moving picture audience, whilst satisfying the dramatic and narrative needs of the film.

On the concluding position that sound often occupies in the film post-production process, Sider notes:

> Meanwhile, sound in film remains, as it has for decades, a more or less technical exercise tacked on to the end of post-production. (Sider, 2003, p.5)

It's generally during the final mixing stage – the commitment of aural artistic intent and usually the last creative production process in the arc of producing a feature film – that the creative work of the Sound Designer and Re-recording Mixer first comes under close scrutiny from the film's Director, as well as other stakeholders; all of whom are engaged earnestly in the endeavour of delivering a finished product to the movie's distributor, by a specified date (which by this stage, is usually looming).

Walter Murch, for whom many accredit the original title and use of the term *Sound Designer*[6](Sider, 2003) describes his *modus operandi* as being to:

> balance the original dreams of the Director, the needs of the studio, my own hunches about things, and the voices of everyone else working on the film. (Loewinger, 1998)

In an expensive-by-the-hour audio post-production mixing theatre, time, tempers and patience can run short; and this can be less than conducive to making creative decisions. What makes this an even more dispiriting destination to arrive at, is that the sound department's capability to support and enhance a film's vision, aspirations or storyline – through the medium of the soundtrack – is often painfully misunderstood by the non-sound professionals involved.

As Sider observes:

> Whether it is credited as 'Sound Design' or 'Sound Editing', sound for film is still largely considered a technical domain only fully understood by a film's sound department. Young filmmakers adopt this attitude early in their training in film-production classes. Rarely is sound included in theoretical analyses of dramas or documentaries. Possibly because of its omnipresence, sound is rarely considered in film pre-production. (Sider, 2003, p.7)

From experience, double Academy Award-winning Sound Designer Randy Thom (*The Right Stuff* (1983), *The Incredibles* (2004)) endorses this very point:

> At the beginning of a project the Director will probably tell you, the Sound Designer/Composer, that sound is extremely important to the film. He's being sincere, but what he really means is something like the following: The sound in this movie has to be great. I don't have time to put much energy into it myself, and I didn't learn much about the creative aspect of film sound in film school where I got the impression that sound work is a series of boring technical operations you don't understand unless you're a physics major, but I'm hiring you because you're supposed to be a genius. You're so brilliant (I hope) and have access to such hi-tech gadgets (I pray) that the track will miraculously, without [the] benefit of actual collaboration, (no time for figuring out what it might mean to let sound collaborate) fill the gaps left by the visual effects and dialogue. (Thom, 1998)

It is my intention to directly address this kind of attitude towards sound for moving pictures.

In my attempts to populate a dialogue that is useful for students, academics and practitioners, the book's three parts are made up of:

- Part 1 – *Constructing the soundtrack*: which explores the ways in which a Sound Designer and a Re-recording Mixer's expertise and underpinning knowledge might be extended, through a better understanding of what emotions they might reasonably be able to induce in audiences, with soundtrack composition and emphasis in mind.
- Part 2 – *Deconstructing the soundtrack*: this provides an analytical breakdown of three major feature film soundtracks; and also investigates in depth three practice-based studies, where acting as both the Sound Designer and Re-recording Mixer, I have used the Four Sound Areas framework to evoke and enhance emotion in an audience, through the balancing of the soundtrack's mix.
- Part 3 – *Communicating the soundtrack*: this presents a new way for film creatives to communicate emotional intent for a soundtrack, through the concept of colour-coding the Four Sound Areas framework and a Mixing visual aid.

Overall, this book is an exploration of the soundtrack, and a journey into the act of audio mixing; it is not specifically concerned with the aural content that is created and mixed.

1.3 The question posed

If there were one single, overarching question that this work aims to address, it would be:

Can the emotional impact of a film scene be affected by balancing a soundtrack's sound design elements in specific ways?

And the approach I have taken to answer that question is several-fold, including:

- Exploring existing professional and academic literature in the field of emotions and sound design.
- Describing the Four Sound Areas framework theory to encompasses sound design and emotion, describing its use in my own sound design work, as well as suggesting its presence and validity that may be found in other sound practitioner's work.
- Examples of feature film analysis, using the proposed framework.
- Examples of my own feature film and television work that used the framework as the basis for soundtrack creation.
- Reflecting on how effective my own soundtracks were in communicating their intended emotions by using the framework.
- Bringing together any new understanding that reflects the applicability, limitations and scope of the proposed framework.
- Describing a new way of visually communicating sound design.
- Summarizing whether this work answers the original question posed.

1.4 The original content and terms used in this book

The soundtracks of the three programmes used for detailed analysis in Part 2 of this book may be considered for academic purposes as being examples of 'practice-based research'. Each is my own original piece of work, created within a professional production and post-production environment, and each utilized the Four Sound Areas framework for the purpose of producing emotive sound design. They were all commissioned for commercial purposes and the theatrical release dates of the two feature films – *The Craftsman* and *Here and Now* – are given in Appendix 1.

1.5 An introduction to the Four Sound Areas in use

In the short animated film, *Jack – Safe@Last*, a practical example is presented of the Four Sound Areas in use, showing how their balancing was designed to evoke predetermined feelings towards the subject matter. The intended result is that a listening-viewer's emotional response to the film will result in a greater engagement with the story than narration and music alone would have achieved.

The mixed soundtrack has extra sounds that sit around the 'untreated' narration – there is no reverberation used on the voice, for instance (which could be used to suggest room size) nor is a contrast provided when the narrator moves from the indoors to outdoors locations (were either of these approaches used, they would be a component of the Spatial sound area). Overall, the full mix was designed to add greater emotional impact to the film's central message, than if the voice was alone presented.

In acting as both the Sound Designer and Re-recording Mixer for this short film, I aim to demonstrate how both the *nature* of the sounds chosen, and their relative *balance* to each other, contribute to the evoking of desired emotions in the listening-viewer.

1.5.1 The starting point

The children's charity *Safe@Last* provides support, shelter and preventative pro-grammes for young runaways in Yorkshire; and the organization commissioned three short animated films to help increase the impact of their preventative work in schools, by letting young people know about their services. The narration for each of the central characters was voiced by real people telling their own story, that the *Safe@Last* charity has helped. This exercise looks at one of those films.

The animation for the three short films was created around an amateur sound recording (the narration), which due to the required anonymity of the children, was recorded by the charity itself.

It is normal practice for animation sequences to be initially arranged on a story-board and then for the animation to commence, timed to pre-recorded dialogue, to ensure that the animators can achieve close lip-synchronization for any on-screen characters; but due to time, budget and inexperience issues, the pre-recorded dialogue was not looked at by a sound editor. Instead the character recordings received a loose dialogue edit by the picture editor, who then went on to animate the images; and so with the poor lip-synch established, a certain *fait accompli* was delivered when the audio Open Media Format (OMF) file and accompanying QuickTime movie was unwrapped in the sound studio, preventing a more seam-lessly edited dialogue track.

1.5.2 The Sound Design brief

The simplistic, animated pictures that accompany the narrative show the loca-tions, environments and situations that the central character, Jack, refers to; and the story takes the listening-viewer on a journey: from the unhappy position Jack found himself in at home with his mother and step-father, through running away for the first time, being found and placed unhappily with a temporary foster-fam-ily, running away for a second time, then making contact with *Safe@Last* through their web-chat facility; meeting up with the *Safe@Last* project worker who was able to gain Jack's trust and then help place him with a permanent foster family, who were able to make things better for Jack – by listening and making him feel valued, and subsequently by taking care of him on a more permanent basis.

The arc of the story has four main story-points:

1) Jack's life at home and his compunction to run away.
2) Jack finding *Safe@Last* and settling with a new family.
3) The charity's messages presented as graphics to young viewers in need of help.

4) Jack's endorsement of *Safe@Last* to help anyone who might be facing what he has been through.

The soundtrack is designed to convey both empathy and sympathy with Jack, through the 'fitting' – by extensive editing – of the supplied musical track, and by the thoughtful selection of Narrative, Abstract and Temporal sounds that are carefully balanced to bring the animated images to life, and to emphasize and reinforce the nature of the situations he has faced in his short life.

The sounds were also chosen to provide an engaging, immersive and credible audio backdrop to the story being told. The soundtrack is designed to take the listening-viewer on a journey: one that has the intention to elicit negative human emotions such as sadness, fear, anger, surprise and disgust, as well as guilt, and tension, before arriving at a much more optimistic, hopeful and positive point of view by the time the 'call to action' arrives at the conclusion of the film.

1.5.3 Utilizing the Four Sound Areas

00:00–00:15 Scene 1 pictures: A street scene at night, and a row of three terraced houses. The sky is starlit, with little cloud and a full moon. It looks a cold, clear night. Traffic passes through shot to wipe and reveal new captions; it is a main road. The two outer houses have pretty, open curtains and are well lit inside; but the middle house is dark, with curtains drawn, although a downstairs room shows a chink of dim light. A caption reads: 'Ever thought about running away from home?'

Scene 1 sound: The film opens on Abstract sound: an atmosphere track created by mixing light and distant city traffic with wind and creaking branches, with synchronized spot effects (lorry passing left to right, two cars from either side and a single car passing right to left); and on Temporal sound: music in the form of a low, mid-tempo acoustic guitar and piano, playing a repeating figure. *Intended listening-viewer emotion*: anxiety.

00:15–00:42 Scene 2 pictures: The camera tracks across the street and through the darkened downstairs window to reveal a young schoolboy, Jack, still in his school uniform, in a stark living room that is illuminated by a single bare bulb which is gently swaying from the draughts within the house. A baby's crib is just behind him and a teddy has fallen out of the crib and rests against the leg. The teddy is worn and used, and clearly not new for this baby.

Scene 2 sound: The Temporal sound of the music continues at low-level and at the same tempo; the Abstract sound consists of the same wind and creaking branches, but now at a lower-level than the outside shot; Jack's Narrative sound narration tells of how his step-Father and Mother would leave him alone at night with his baby sister, necessitating him to take care of her needs; until they returned home from the pub and had drunken arguments with him. A baby begins to cry (also Narrative sound) as Jack talks about looking after her. *Intended listening-viewer emotion*: sadness.

00:42–00:50 Scene 3 pictures: Jack is alone in a pool of dim light.

Scene 3 sound: Jack continues his story in the Narrative sound area. The Abstract sounds are the continuing wind and creaking branches and the Temporal sound is the low-level music. *Intended listening-viewer emotion*: unhappiness.

00:50–01:04 Scene 4 pictures: Jack is standing on the street, a main road, and it is raining. A car passes by, left to right.

The Narrative sound of Jack's monologue continues, as does the Temporal sound of the music. The Abstract sounds of the wind and creaking branches have raised again in level and are mixed with the sound of rain falling on grass and concrete. A spot effect of a car passing from left to right is also added to the Abstract sound area. *Intended listening-viewer emotion:* insecurity.

01:04–01:20 Scene 5 pictures: A family sitting room, with a hamster cage in the corner. Jack is alongside the temporary foster parents he has been placed with.

Scene 5 sound: The Narrative sound is Jack's narration, Abstract sound is the continuing, low-level wind and creaking branches and Temporal sound is the repeating low-level music. *Intended listening-viewer emotion*: isolation.

01:20–01:26 Scene 6 pictures: Jack is back on the street.

Scene 6 sound: Narrative sound is provided by Jack's narration and Temporal sound is the same repeating phrases of music. Abstract sounds are the increased wind and creaking branches, mixed with a city traffic atmosphere. A distant Police siren is timed appropriately to the narration (the siren being codified Narrative sound); the siren a signal of the danger Jack is in by being back on the street again. *Intended listening-viewer emotion*: apprehension.

01:26–01:43 Scene 7 pictures: The shot widens on the street where Jack is standing to reveal that he is outside a Public Library. The camera tracks into the Library and we see Jack sitting in front of a computer. On the monitor screen is the *Safe@Last* logo.

Scene 7 sound: Jack's Narrative sound narration continues, joined by the Narrative sound of keyboard strokes, reinforcing the notion of communication, and indicate that Jack is now directly in touch with *Safe@Last*. Intentionally, there are no associated sounds with the text replies appearing on the PC screen in reply, as hearing on-screen text making sounds is an unrealistic use of artistic licence. The Abstract sound leitmotif of wind and creaking branches progressively fades; this change in the sound design associated with a positive change in Jack's fortunes. Accordingly, the music is edited to begin moving away from its unresolved, initial cycle and is preparing to change tempo and become more upbeat. *Intended listening-viewer emotion*: hope.

01:43–02:28 Scene 8 pictures: Jack is on a climbing frame in a children's play area, alongside his *Safe@Last* caseworker. The shot widens to reveal that his new foster-parents are there too.

Scene 8 sound: The Narrative sound continues with Jack's narration; and the Abstract sound is a mix of residential traffic and birdsong, with the sounds of children happily playing off-screen. As the shot reveals Jack's new foster-parents, the level of birdsong increases in the mix, suggesting and reinforcing the notion of Jack's safety and well-being. The Temporal sound is the music, which is edited carefully for the positive tempo and instrumentation change to coincide with the next scene. *Intended listening-viewer emotion*: optimism.

02:28–02:34 Scene 9 pictures: Caption – 'We're here for you whenever you want, however you want'.

Scene 9 sound: All Narrative and Abstract sound is removed, leaving just the Temporal sound of the music score. The scene change coincides with the swelling of the music and the punctuation provided by percussion and bass being introduced to the score, bringing about a mood-change in the music: from the tentative, cautious and unresolved opening figures to the up-beat, positive and confident bars that go on to accompany scenes 10, 11 and 12. Whilst not suggesting that Temporal sound alone is responsible for the feeling of happiness, the musical 'surge' that is pushed in the mix is heightened by removing any hint of distraction from other 'real-world' sounds in the Narrative or Abstract sound areas; in effect, de-coupling the 'emotion' of the moment from the 'reality' of the scene. *Intended listening-viewer emotion*: relief.

02:34–02:39 Scene 10 pictures: Caption – 'Call us free and in confidence on 0800 335 SAFE'

Scene 10 sound: As scene 9

02:39–02:43 Scene 11 pictures: Caption – 'Text SAFE and your message to 60777'

Scene 11 sound: As scene 9

02:43–02:47 Scene 12 pictures: Caption – 'Or go to www.safeatlast.co.uk'

Scene 12 sound: As scene 9

02:47–02:55 Scene 13 pictures: Jack in frame, talking to the camera. The shot widens to reveal he is standing in front of the house in the *Safe@Last* logo.

Scene 13 sound: The Temporal sound area music track dips to allow Jack to pick up his narration (this is in the Narrative sound area), endorsing the positive experience he has had with *Safe@Last*, and recommending them to others who may be facing the difficulties he encountered. This positive message is underlined by the Abstract sound elements of forest birdsong, intended to represent safety and well-being, and the sound of running water in the form of a small flowing stream, intended to suggest cleansing and renewal. *Intended listening-viewer emotion*: satisfaction.

02:55–03:00 Scene 14 pictures: The *Safe@Last* logo.

Scene 14 sound: The film ends on the Abstract sounds of birds and water and the Temporal sound area of the music track, carefully edited to finish appropriately with the end of the film; the music's last diminished chord suggesting reflectively, but without melancholy, that happily, Jack's story is not entirely finished. *Intended listening-viewer emotion*: triumph.

Sound Design summary

The success of evoking certain emotions in the listening-viewer is dependant not just on the types of sounds selected for inclusion in a soundtrack, but importantly, the degree of their presence in the overall mix, e.g.:

- What levels are the Abstract and Temporal sounds set to when sitting alongside the Narrative sound area's speech?
- To what level does the Temporal sound area's music rise at times of significance in the timeline?
- Which sound areas override others for emotional impact, or artistic effect, at specific moments?

As discussed earlier in this chapter, the decision over the *type* of sounds used in a soundtrack is usually that of the Sound Designer; whilst their relative *balance* is usually set by the Re-recording Mixer. (However, in many instances, as in this film, the same person fulfils both roles.)

The questions surrounding how the *amounts* of Narrative, Abstract, Temporal and Spatial areas change in a mix, the way that they are continually being rebalanced by the Re-recording Mixer against each other, and the decisions behind which sound area has precedence over any other at a particular moment of emotional significance, are key to determining if, how and by how much the listening-viewers emotions may be successfully directed.

Perhaps it is the *type* of sound that appeals to the *intellect*, whereas the *balance* of the sound, the more visceral of the two conditions, appeals to the listening-viewer's *instinct*.

The balance of the mix itself may also be thought of in visual terms: whilst the *type* of sounds heard frame a scene, it is the mix *balance* that provides the focus.

In summary, the Narrative sound elements consist of Jack's narration, a baby crying and the codified sounds of a Police siren and keyboard strokes.

The Abstract sound elements chosen consist of an atmospheric bed of wind and creaking branches, suggesting restlessness and unease within an unsettled situation; which is replaced by birdsong and the sound of happy children playing (their voices indistinct and therefore not used in a Narrative sound sense). The final, positive message of the film is reinforced by the inclusion of the sounds of birds and water, suggested as signs of safety and renewal, balanced against the Temporal sound music and included in the Spatial area of the surround-channels.

The Temporal sound element consists entirely of music, extensively edited from its original form, to initially repeat an early, simple figure that suggests both uncertainty and little progress, as the music remains unresolved. The first change of tempo provides the resolve and this temporal change heralds a more upbeat, optimistic mood that coincides with the turning point of Jack's story: the breaking of the vicious circle he has been trapped by, up to this point. The final 'call to action' message has the music edited to coincide with its composed conclusion and the full instrumentation and musical arrangement.

The Spatial area is not explored to any extent.

Throughout the mix, Jack's narration (Narrative sound) takes precedence over all the other sound areas.

The Abstract sound of passing cars and the Temporal sound of music in scene 1 (intended emotion: *anxiety*) are equally weighted, but in scene 2 (intended emotion: *sadness*), the sound of a baby crying is allowed prominence over the music and the other Abstract sounds of wind and creaking branches, which are mixed to be present, but not prominent.

In scene 3 (intended emotion: *unhappiness*), the baby has stopped crying and the narration is firmly foreground. The music occupies the middle ground, whilst the Abstract sound is in the background.

In scene 4 (intended emotion: *insecurity*), the Abstract and Temporal sound areas have swapped places with each other; the emphasis underneath the narration being on the Abstract atmosphere track, made more dense by the sound of rain falling on both grass and on concrete, and the spot effect of a car passing from one side of the screen to the other.

In scene 5 (intended emotion: *isolation*), Jack is back indoors, with the layers returning to: foreground – Narrative narration; middle ground – Temporal music; and background – the Abstract atmosphere track of wind and creaking branches.

In scene 6 (intended emotion: *apprehension*), the Narrative sound of a Police siren is timed to be noticed between the words of Jack's Narrative narration, but allowing the narration to be heard unimpeded; the Abstract sound atmosphere of city traffic overriding the continuing creaking branches and wind to suggest busy, anonymous streets and mixed to sit alongside the Temporal music.

In scene 7 (intended emotion: *hope*), there are several distinct layers of sound, the first two being Narrative (Jack's voice above the keyboard strokes, the speech higher in level than the keyboard); underneath this foreground is the Temporal music, replacing in the middle ground the fading creaking branches and wind atmosphere track, as it moves progressively to the background.

In scene 8 (intended emotion: *optimism*), underneath the narration, but prominently mixed alongside the music, the Abstract sound of the children playing out-of-doors is less of a suggestion than a statement; the camera moves to a wide-shot, revealing Jack's new foster parents, coinciding with the introduction of birdsong, which is noticeably added to the mix. The Narrative and Temporal sounds have remained at a constant level whilst the Abstract sound has changed in this scene; but once the birdsong is established, it is then the turn of the Temporal sound to begin to rise as the scene draws to a close; the Narrative voice pausing and the Abstract sounds fading out completely – music being the only sound for scenes 9–13 (intended emotion: *relief*).

In scene 13 (intended emotion: *satisfaction*), the Temporal music is dipped as the narration re-starts, to allow prominence once again to the Narrative sound of Jack's voice. Initially, to establish the presence of the flowing water and birdsong in the listening-viewers' consciousness, the music dips momentarily below the level of the water and birds, before rising again to sit closely, but not quite alongside this Abstract area.

In scene 14 (intended emotion: *triumph*), and with the narration complete, the music takes precedence over the still noticeable water and birds; the increase in level of the Temporal music over the Abstract atmosphere track allowing space for the development and fulfilment of the music's conclusion.

1.6 Discussion points

- What are the roles and responsibilities of the sound department? (Consider 'production' as well as 'post-production' Sound.)
- What is the purpose of sound design and who has the final say?
- At what stage does sound design begin? (Consider the pre-production, production, and post-production stages.)

1.7 Practical exercises

- Discuss and explore the basic capabilities of a DAW.
- Explore the 14 ways (or more?) that an audio clip might be manipulated.

Notes

1 The term 'Sound Designer' was originally a US term, with the title 'Supervising Sound Editor' often used instead in the UK. This text will use the term Sound Designer throughout.

2 The term 'Re-recording Mixer' is similarly exchangeable with the UK title 'Dubbing Mixer'. This text will use the term Re-recording Mixer throughout.

3 Whilst prolific Re-recording Mixer Richard Portman (b. 02-04-1934 – d. 28-01-2017), son of early sound-cinema Re-recording Mixer Clem Portman (*King Kong*, 1933; *Citizen Kane*, 1941; *The Thomas Crown Affair*, 1968) has been described as the first *Hollywood* Mixer to have mixed a feature film single-handedly (Rozett, 1998), this practice was already well established outside of the major studios: Walter Murch solo-mixed many films including *The Rain People* (1969), *THX-1138* (1971) and *The Conversation* (1974) in San Francisco, before teaming up with Mark Berger to mix *The Godfather: Part II* (1974) (Murch, 2020).

4 Mixing 'in-the-box' refers to a workflow where audio assets are stored, replayed and manipulated within the computer programme of a Digital Audio Workstation, as opposed to the older method of sound sources all appearing on individual mixing console channels for adjustment of, e.g. level and equalization. Modern, large track-count feature films often combine these 'old' and 'new' methods by the arrangement of a DAW being coupled to a very large mixing console.

5 This possibly understates the old Re-recording Mixer's adage of 'I need to make 100 decisions, so the Director only has to make 1'.

6 Murray Spivak (b. 06-09-03 – d. 08-05-94) is arguably the first proponent of sound design, famously through his ground-breaking work on *King Kong* (1933) that employed reversed recordings of real animals to suggest the monstrous nature of the creature Kong. He won an Academy Award for Sound Recording on *Hello Dolly* (1969) and was nominated for an Oscar for his work on *Tora! Tora! Tora!* (1970).

Part 1

Constructing the soundtrack

You are not the only one who is confused on where to start. We have to be brave, we have to be strong, and we have to believe in ourselves.

Malala Yousafza

2 Emotion in sound design

Introduction

This chapter looks at notable examples of current professional and academic literature that are relevant to this topic, and to the increasing interest in the study of sound and emotion. This chapter also clarifies the terminology used, discusses to what extent the relationships of music and emotions – and speech and emotions – are relevant to this book, and looks at existing sound-related theoretical structures.

2.1 Defining the nature of emotions

The original motivation for this book came from my desire as a professional Sound Designer to investigate and understand quite what the 'emotional' element of an audience's reaction to moving picture sound design, and soundtrack balancing, might be; and then, if it was possible to shed light on what that is, to look at whether reproducible techniques might be employed by fellow Sound Designers and Re-recording Mixers, to elicit target emotions in an audience.

Clearly, audiences do have emotional reactions to movie soundtracks – there are obvious, outward signs when a film makes an audience laugh out loud in a theatre, or even cry; and many of us will have first-hand experience of what it is to feel happy, fearful or uncomfortable whilst watching (and listening) to a film.

However, after reading works relevant to my investigation, it soon became apparent that it was important for me to determine whether the response of an audience to soundtrack stimuli could be described consistently; because there appeared to be a distinct commingling of the terms audience *emotion* and audience *affect*. And in seeking an answer to the question of which is the most appropriate term between *emotion* and *affect*, there seemed to be many examples of misunderstanding, or even misappropriation, of the two terms.

It is not helped by the fact that any attempt to provide a simple, clear-cut clarification of the difference between *affect* and *emotion* is somewhat challenging;

not least of all because language and concepts become increasingly abstract the deeper one delves into specialist works. However, Massumi proposes that:

> Affect is 'unformed and unstructured,' and it is always prior to and/or outside of conscious awareness.
>
> (Shouse, 2005)

Whilst in his essay 'Why Emotions Are Never Unconscious', Clore proposes that:

> emotions that are felt cannot be unconscious.
>
> (1994, p. 285)

Therefore it may be reasonable to suggest that considering affect as an unconscious process, whilst regarding emotion as a conscious one, could be a good place to start in differentiating the two terms; at least for the purpose of this conversation – so, for instance, the bodily response that arises due to the threat of an oncoming vehicle, or being caught in a tsunami, could be considered as an example of *affect*; whereas crying in sympathy with an on-screen character's situation would be seen as an example of *emotion*.

But Shouse, for one, is specific about the difference between the terms emotion and affect:

> Although feeling and affect are routinely used interchangeably, it is important not to confuse affect with feelings and emotions. As Brian Massumi's definition of affect – in his introduction to Deleuze and Guattari's *A Thousand Plateaus* – makes clear, affect is not a personal feeling. Feelings are *personal* and *biographical*, emotions are *social*, and affects are *prepersonal*. (Shouse, 2005)

And then, by considering an aspect of the work presented by Deleuze and Guattari themselves, specifically their 'autonomy of affect' theory, which proposes that affect is independent of the bodily mode through which an emotion is made visible (Schrimshaw, 2013), it seemed to be incongruous, particularly as far as the topic of this study is concerned, to elevate the impersonal concept of *affect* over the personal and social factors that constitute a cinema-viewing experience, and more readily align with the term *emotion*.

Another clarification of the two terms is provided by Lisa Feldman Barrett, writing an endnote to Chapter 4 of her book *How Emotions are Made: The Secret Life of the Brain*:

> Many scientists use the word 'affect' when really, they mean emotion. They're trying to talk about emotion cautiously, in a non-partisan way, without taking sides in any debate. As a result, in the science of emotion, the word 'affect' can sometimes mean anything emotional. This is unfortunate because

affect is not specific to emotion; it is a feature of consciousness. (Feldman Barrett, 2017)

Furthermore, Shaviro is emphatic about what is primarily engaging an audience:

Reading a novel, hearing a piece of music, or watching a movie is an emotional experience first of all. Cognition and judgment only come about later, if at all. (Shaviro, 2016)

And so, throughout this book, it is proposed that the context for the work undertaken by a Sound Designer and Re-recording Mixer most appropriately lies within the boundaries of influencing audience *emotion*.

The challenge of defining what constitutes an emotion remains, however. As Kathrin Knautz observes, whilst it may be straightforward to determine our own, because of the difficulty in defining emotions, some researchers resort to formulating a definition by instead looking at the features of emotions. (Knautz, 2012)

Fehr and Russell comment on this conundrum:

Everyone knows what an emotion is, until one is asked to give a definition. Then, it seems, no one knows.

(1984, p. 464)

In the introduction to his book *From Passions to Emotions*, Dixon (2003) suggests that the rise in academic work in a range of fields concerned with the emotions is a modern trend, one that is in direct contrast to the preoccupation with intellect and reason of earlier studies. Furthermore, he feels that this is no bad thing:

Being in touch with one's emotions is an unquestioned good. (Dixon, 2003, p.1)

Through his research on *Pan-cultural recognition of emotional expressions* (Ekman et al., 1969) and his subsequent work *Basic Emotions* (Ekman, 1999), Ekman suggests that six fundamental emotions exist in all human beings: happiness, sadness, fear, anger, surprise and disgust.

Plutchik (2001) broadly agrees with Ekman, but further develops the categories by creating a wheel of eight opposing emotions, where positive emotions are counterpointed by equal and opposite negative states: joy versus sadness; anger versus fear; trust versus disgust; and surprise versus anticipation.

From Ekman and Plutchik's definitions, Antonio Damasio (2000) further suggests that more complex emotional states can arise: such as embarrassment, jealousy, guilt, or pride (sometimes referred to as *social* emotions), or well-being, malaise, calm, tension (*background* emotions). As one of the world's leading experts on the neurophysiology of emotions, Damasio summarizes the fact that without exception men and women of all ages, social and educational backgrounds are subject to emotions; he also refers to the way different sounds evoke emotion:

> Human emotion is not just about sexual pleasures or fear of snakes. It is also about the horror of witnessing suffering and about the satisfaction of seeing justice served; about our delight at the sensual smile of Jeanne Moreau or the thick beauty of words and ideas in Shakespeare's verse; about the world-weary voice of Dietrich Fischer-Dieskau singing Bach's *Ich habe genung* and the simultaneously earthly and otherworldly phrasings of Maria João Pires playing any Mozart, any Schubert; and about the harmony that Einstein sought in the structure of an equation. In fact, fine human emotion is even triggered by cheap music and cheap movies, the power of which should never be underestimated. (Damasio, 2000, pp. 35–36)

The first studies of emotion with regard to sound were related to music and came in the late nineteenth century, coinciding with psychology becoming an independent discipline around 1897; although the early peak in studies was seen sometime later, in the 1930s and 1940s (Juslin and Sloboda, 2010).

Today, a multidisciplinary approach pervades the field of emotion in music, and although there is not yet unanimous agreement on whether there are uniquely musical emotions, or whether the nature of these emotions is basic or complex, the field of emotion in music is steadily advancing (Ibid.).

Jenefer Robinson articulates the complexity that the analysis of music and emotions can produce:

> the sighing figure is heard as a sigh of misery (a vocal expression), a syncopated rhythm is heard as an agitated heart (autonomic activity), a change from tonic minor to parallel major is heard as a change of viewpoint (a cognitive evaluation) on the situation from unhappiness to happiness, or unease to serenity, and given the close connection between the two keys and the fact that the melody remains largely the same, we readily hear the evaluation as ambiguous or as shifting: the situation can be seen as both positive and negative. […] Overall, we may hear the piece as moving from grief and anguish to serene resignation, all of which are cognitively complex emotions. (Robinson, 2005, p. 320)

However, Juslin and Sloboda broaden the perspective of the way sound can evoke emotion from that of a purely music-based discussion, by suggesting that it is now recognized that a significant proportion of our day-to-day emotions are evoked by cultural products other than music; and therefore designers should be mindful of emotion in the products and interfaces that they design, in order to make them richer and challenging to the user (Juslin and Sloboda, 2010).

From the advent of the medium, moving picture producers have described and promoted their films by describing the emotions that the audience is intended to feel when they watch them (e.g. horror, romantic-comedy, or mystery-thriller).

So, it is reasonable to suggest that audiences have proven themselves to not only being susceptible to, but even desirous of having their emotions evoked in a movie theatre.

Holland, in *Literature and the Brain* (2009), writes on our emotional response to literary work, of which cinema is an important part:

> The brain's tricks become clearer at the movies.
>
> (Holland, 2009, p. 2)

Clearly then, it is important to consider what might be happening to an audience as they watch a movie.

Holland proposes that a well-designed soundtrack is instrumental in engaging and enveloping a viewing audience, and a listening-viewer absorbed by on-screen activities forgets their own body and its immediate surroundings, enabling them to be transported to all kinds of otherwise improbable locations and situations. Central to Holland's line of reasoning is that an emotion is a call to action, or a disposition to act; yet when we sit in the cinema and have our emotions evoked through the sound and pictures we are viewing, we remain seated. This, he suggests, is due to a unique contract with the work. Even though we are figuratively transported by our emotions towards a certain state of mind, we identify that it is the circumstances of the on-screen activity or character that has aroused these feelings within us, and it is not a direct consequence of us being in the represented situation (Holland, 2009).

Because most bodily responses brought about by emotions are visible to others, they in turn bring about 'mirroring' in the viewer. Humans tend to respond to emotional expressions they see with similar emotions themselves; and as early as 1890, Darwin noted that emotions communicate in this fashion (Darwin, 1890).

But Holland suggests that since it is a mirroring process at work, the impulse to act on the emotion is inhibited: i.e. whilst watching certain actions, motor regions of the brain experience an *impulse* to act (the mirroring). However, the brain *inhibits* this musculoskeletal expression through a process called the 'inverted mirror response'; more fully described by Marco Iacoboni (2008) in his work on 'super mirror neurons' (Holland, 2009).

For Holland though, mirroring is not the complete picture of a fuller, immersive and emotional involvement with an on-screen subject. Our own past experiences of circumstances like the viewed events are also powerfully evoked; and he states:

> We bring to bear on what we now see, some feeling or experience from our own past. And my bringing my own past to bear on the here and now of tragedy makes me feel it all the more strongly. (Holland, 2009, p. 72)

Richard Gerrig includes this in what he calls a 'participatory response', and he notes how it can enrich and intensify one's 'emotional experience' (Holland, 2009, quoting Gerrig, 1996).

It is evident then that sound triggering or affecting human emotions is not just limited to music; other sounds too can contribute to this process. Certainly, some of the wider range of emotional stimulation that Damasio describes sits comfortably within the remit of the audio post-production stages of filmed stories, or televised drama. Juslin and Sloboda's comments also suggest that there is both scope and a basis for the thoughtful use of soundtrack elements to evoke emotional responses within a listening-viewer; and Holland's description of how audiences engage with what they see on-screen would seem to further support this proposition.

2.2 The relevance of speech and emotions research, and music and emotions research, to this study

Whilst there is little research yet dealing specifically with moving picture sound design and emotions, there is a substantial body of research concerning both speech and emotions (e.g. Banse & Scherer, 1996; Cowie, 2000; Pereira, 2000) and music and emotions (e.g. Hunter & Schellenberg, 2010; Juslin & Sloboda, 2010; Swaminathan & Schellenberg, 2015).

Speech and music are two key elements of the compound that constitutes a moving picture soundtrack; and both contribute greatly to the viewing experience of movie audiences, not only by virtue of their *expressing* of emotion, but also by their being capable of *inducing* emotion in listening-viewers.

Three aspects of *speech and emotions* research are particularly relevant in this study.

First and foremost, both speech and a film's soundtrack are designed to communicate with an audience. A film soundtrack, intended as a compound of speech, sound effects and music, not only has the ability to be as literal as speech in portraying emotions (indeed it contains speech and therefore a character can utter words such as "I feel sad", telling the audience explicitly what emotion is at play), it can also be more so than a musical score alone might.

However, it is important to make clear that this statement is not intended to diminish the importance of music in movies. Far from it, music is a powerful emotional tool, particularly when skilfully deployed within a film soundtrack (e.g. Damasio, 2000).

Many movies are most memorable precisely for their featured musical interludes,[1] which create iconic snapshots that go on to define a production, long after the film's fuller storyline has left the consciousness of audiences; e.g. *Tiny Dancer* (Comp. Elton John/Bernie Taupin) in *Almost Famous* (2000) (Dir. Cameron Crowe/Sound Designer Mike Wilhoit), *Bohemian Rhapsody* (Comp. Freddie Mercury) in *Wayne's World* (1992) (Dir. Penelope Spheeris/Sound Designer John Benson) or *Always Look On The Bright Side of Life* (Comp. Eric Idle) in *Life of Brian* (1979) (Dir. Terry Jones/Re-recording Mixer Hugh Strain) to name but three of a long, 90-years-plus list, that began with *The Jazz Singer* (1927) (Dir. Alan Crosland/Sound Engineer Nathan Levinson), the film widely considered to be the first commercial 'talkie'.[2]

But if songs or arias with a text are discounted, it is reasonable to argue that a music score is less directly meaningful, and overall, it is more abstract than literal in its nature.

As an aside to this immediate point, but nonetheless still highly relevant to the way music is used in movies, there is also the constant consideration by the Re-recording Mixer that music has the ability to emotionally overwhelm a soundtrack, particularly if its application is not judiciously metered and carefully balanced with the other mix elements.[3] As Sider suggests:

> Rather than allow the audience to come to their own conclusions the music presses an emotional button that tells the audience what to feel, overriding the words and thoughts of the film's characters. (Sider, 2003, p. 9)

Tarkovsky would seem to go further:

> Above all, I feel that the sounds of this world are so beautiful in themselves that if only we could learn to listen to them properly, cinema would have no need of music at all. (Tarkovsky, 1987, p. 162)

So whilst this book looks carefully at the interplay between dialogue and sound effects, a relationship to which music also makes a conspicuous contribution, music in this study is treated respectfully for its emotional power in its own right; but from a Re-recording Mixer's perspective, music is but one of the sounds that require balancing.

Because all sounds – not just music – can be emotionally important in a movie (e.g. a single gunshot suddenly featured in a scene that had only music playing will immediately draw the listener's attention away from the music) and whilst a sound may be interpreted in several ways, often depending on the context it is heard in, all sounds in this study are referred to, considered as, or classified by, their *primary emotional function* or purpose in the soundtrack.

And so, through the combination of all these sounds, the relative proportions of which are solely determined by the Re-recording Mixer during the act of pre-mixing and final mixing, the underlying meaning of the soundtrack is revealed.

Secondly, when considering the soundtrack and the way it forms part of an audio-visual work, there are comparisons that may be drawn between the Re-recording Mixer's mix-balancing with an emotional intent in mind, and the way that everyday speech is used to convey emotion. In speech, the *meanings* of words are quite fixed within a language, yet the actual *emphasis* of the words being spoken can be quite fluid due to inflection, tonality or accent.

The emphasis on words plays an important role in inducing different emotions in the listener. For example, I might say the words 'I'm really sad' in a helpless sounding way, or in a sarcastic sounding way. The words are the same and indicate an emotion, but the *sound* of the words will determine the emotion that the listener will perceive.

So too in a movie, where the words of dialogue that the characters use may on their own have clear meaning for the plot and storyline; yet when balanced amongst other mix elements in the soundtrack, what results is a listening experience that is emotionally richer for the other sound elements that have been placed carefully around the speech.

Additionally, the visual elements of a film (the acting, editing, lighting, grading, composition, etc.) can powerfully portray a particular emotional direction (similarly to how the meaning of words do in speech). But the soundtrack, and the balancing of its elements by the Re-recording Mixer, can shift the emotional direction of the overall experience.

This is similar to how the changes in the prosodic patterns that naturally exist in speech produce emotional shifts: e.g. the tendency to speak unwittingly loud when gleeful, or in a higher than usual pitch when greeting a sexually attractive person (Bachorowski, 1999); and this is described in other research studies of listeners inferring emotion from vocal cues (see Frick, 1985; Graham, San Juan & Khu, 2016; van Bezooijen, 1984 to name but a few).

In an audio-visual piece of work with emotional meanings already suggested through the visuals, or through words and other selected sounds, variations in emotional meaning can also be produced by manipulating the mix *balance* of the sound track; which is similar to how natural variations in pitch, loudness, tempo and rhythm do in speech.

2.3   Hearing the soundtrack

In *Listening*, the opening chapter of social theorist and writer Jacques Attali's work *Noise: The Political Economy of Music* (1985), the author attaches a much greater importance to the act of listening than that often attributed to the purely cinematic act of audition, or the emotional effect a soundtrack may evoke:

> For twenty-five centuries, Western knowledge has tried to look upon the world. It has failed to understand that the world is not for the beholding. It is for hearing. It is not legible, but audible. (Attali, 1985, p. 3)

Which implies that sound itself carries a quality, or set of qualities, that can not only inform a cinema audience, but also impart meaning on what they are seeing; which in turn relates to the assertions of Holland (2009) and accords with my notion that (especially) within narrative filmmaking, a significant responsibility is capable of being borne by the soundtrack to fully engage and emote an audience.

In his essay *Art in Noise*, Mark Ward suggests that:

> it is unlikely one may have a meaningful narrative experience without it also being an emotional one. (Ward, 2015, p. 158)

Ward also argues against the primacy of speech and music in the traditional process of soundtrack dissection, instead elevating what might be termed as environmental sound, or sound effects, to a status at least equal to dialogue and score (Ward, 2015). This also implies that these fuller soundtracks require careful balancing by the Re-recording Mixer:

> Sound design […] is considered to be a process by which many sound frag-
> ments are created, selected, organised, and blended into a unified, coherent,
> and immersive auditory image. (Ward, 2015, p. 161)

Ward then goes on to make three key assumptions:

i) Cinema is not a visual medium, but multimodal: what is cinematic about cinema is moving imagery, not moving pictures. (Ward, 2015, p. 158)
ii) Sound can modify visual perception: sound design through careful crafting, may steer and deflect the eye's passage across a screen, or draw the eye to some objects but disregard others. (Ward, 2015, p. 159)
iii) […] contemporary sound design [is] a playful recombination of auditory and visual fragments, and a heightened manipulation of auditory spatialisation, temporal resolution, and timbre. (Ward, 2015, p.161)

In arguing that the cinema experience is an emotional one, Ward sub-categorizes the construction of a soundtrack into three distinct areas; and his citing of auditory spatialization and temporal resolution directly accord with two of this study's Four Sound Areas, e.g. Spatial and Temporal (which will be more thoroughly described in Chapter 4).

Michel Chion also utilizes a tripartite classification when he describes the way in which soundtrack elements are heard by an audience; and he refers to these three states as *causal*, *semantic* and *reduced listening*.

Causal listening, the most common form of listening mode

> consists of listening to a sound in order to gather information about its cause
> (or source).
>
> (Chion, 1994, p. 25)

Causal listening can condition, or even prepare, the listener by the very nature of the sounds heard – for instance, the sound effect of a dog barking readily recalls the image of a dog in the listener.

Chion goes on to describe how a film soundtrack might manipulate causal listening through its relationship to the pictures; a term he calls *Synchresis*; whereby we are not necessarily listening to the initial causes of the sounds in question, but rather causes that the film has led us to believe in:

> [In] causal listening we do not recognize an individual, or a unique and par-
> ticular item, but rather a category of human, mechanical, or animal cause: an
> adult man's voice, a motorbike engine, the song of a meadowlark. Moreover,
> in still more ambiguous cases far more numerous than one might think, what

we recognize is only the general nature of the sound's cause. (Chion, 1994, p. 27)

Chion describes semantic listening as

that which refers to a code or a language to interpret a message. (Chion, 1994, p. 28)

For Chion, causal and semantic listening can occur simultaneously within a sound sequence:

We hear at once what someone says and how they say it. In a sense, causal listening to a voice is to listening to it semantically, as perception of the handwriting of a written text is to reading it. (Chion, 1994, p. 28)

Chion thirdly suggests that reduced listening refers to the listening mode that focuses on the traits of the very sound itself, independent of its cause and of its meaning:

Reduced listening has the enormous advantage of opening up our ears and sharpening our power of listening [...] The emotional, physical and aesthetic value of a sound is linked not only to the causal explanation we attribute to it but also to its own qualities of timbre and texture, to its own personal vibration. (Chion, 1994, p. 31)

Finally, Chion asserts that natural sounds or noises have become the forgotten or repressed elements within the soundtrack – in practice and in analysis; whilst music has historically been well studied and the spoken voice more recently has found favour for research:

noises, those humble footsoldiers, have remained the outcasts of theory, having been assigned a purely utilitarian and figurative value and consequently neglected.

(Chion, 1994, pp. 144–145)

Another view of separating an audience's listening processes is proposed by Sound Designer and Re-recording Mixer Walter Murch (*American Graffiti*, 1973; *The Conversation*, 1974; *Apocalypse Now*, 1979).[4] He describes a way in which he views the elements of a soundtrack 'positioned' in a virtual spectrum for auditioning; and he suggests that this positioning is instrumental in how the soundtrack is processed in the brain of the listening-viewer.

In his essay 'Dense Clarity, Clear Density', Murch likens the sound design palette to the spectrum of visible colours: from the colour red at one end of the scale, to the colour violet at the other.

Conceptually superimposing sound on to this visual image, he places what he describes as 'Embodied sound' (the clearest example of which is music) at the Red extreme and what he describes as 'Encoded sound' (the clearest example of which is speech) at the Violet extreme.

With these two extremities of speech and music bracketing the available range, all usable sound must therefore fall between them: with almost all sound effects somewhere in the middle – half-way between language and music. Murch considers these sound effects, whilst usually referring to something specific within a soundtrack, not to be as abstract as music, but nonetheless, not to be as universally and immediately understood as spoken language.

Murch goes on to suggest that separate areas of the brain process the different types of audio information, with encoded sound (language) dealt with by the left half of the brain, and embodied sound (music) dealt with by the right hemisphere. He then proposes that by evenly spreading the elements of his mix between the two pillars of the audio-scale, a clearer (even though busier) soundtrack, with a higher mix-element count, can be achieved than a soundtrack in which multiple mix-elements are concentrated in one particular area of the audio sound spectrum.

This left–right duality of the brain, in Murch's opinion, therefore, enables twice as many 'layers' – five – to be achieved in a soundtrack when the type of sound used is spread, for example:

Layer 1: dialogue
Layer 2: music
Layer 3: footsteps (Murch's 'linguistic effects')
Layer 4: musical effects (Murch's 'atmospheric tonalities')
Layer 5: sound effects.

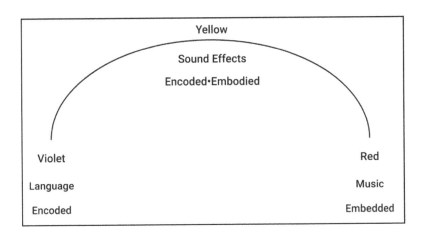

Figure 2.1 Walter Murch's 'Encoded – Embodied' sound spectrum

If, however, you desire two-and-a-half layers of dialogue to be heard simultaneously, elements elsewhere must be sacrificed to retain clarity in this density of dialogue. Murch refers to this phenomenon as his 'Law of two-and-a-half' and this 'rule-of-thumb' is defined by Murch based on his long experience as a Sound Designer, a Re-recording Mixer and sound editor, as well as a picture editor (Murch, 2005).

Ward, Chion and Murch's theories are particularly significant for the central topic of this book as they address issues directly related to soundtrack production and listening-viewers.

2.4 The impact of linking what we hear, to what we see

In her paper 'Making Gamers Cry', Karen Collins suggests that:

> Our emotional and neurophysiological state can be directly affected by what we see: for instance, if we see pain or fear in someone else, we understand this in terms of our own psychophysiological experience of similar pain or fear. For example, neurons that normally fire when a patient is pricked with a needle will also fire when the patient watches another patient being pricked. (Collins, 2011, p. 2)

This highlights the fact that seeing something on-screen can evoke an emotional reaction in the observer's brain through the activity of so-called 'mirror neurons', which are thought to be the main route to human empathy.

Neuroscientist Vilayanur Ramachandran believes that these mirror neurons actually dissolve the barrier between self and others, light-heartedly referring to them as 'Gandhi Neurons' (Ramachandran, 2009).

But what would seem to be highly significant to this investigation into emotions evoked by sound, is what Keysers et al. (2003) described from the research they conducted into monkey mirror neurons, in which they found that the same neurons fired whether an action is performed, seen or simply *heard*:

> By definition, 'mirror neurons' discharge both when a monkey makes a specific action and when it observes another individual making a similar action (Gallese et al. 1996; Rizzolatti et al. 1996). Effective actions for mirror neurons are those in which a [monkey's] hand or mouth interacts with an object. (Keysers et al., 2003, p. 628)

In plain terms:

> These audio-visual mirror neurons respond as if we are experiencing the cause behind the event, when only the sound of the action is presented. In other words, when the monkey hears the sound, the brain responds as if it is also seeing and experiencing the action creating the sound. (Collins, 2011, p. 2)

These results would seem to add credence to the notion that sound alone is a powerful emotional tool that can be put to good use in moving picture production.

This clinically observed reaction to the effect of 'hearing-without-seeing' (which in cinematic rather than laboratory terms could include the practice of 'sound-leading-picture'), is an established sound design technique frequently used to purposely develop the tension of an unsettling event or situation, through the presence of (often) abstract sound effects, whose origination remains for the most part unseen.

However, as the story develops, the Sound Designer in the tracklay, and then the Re-recording Mixer in the mix itself, may consider that what originally were Abstract area sounds, later on contribute to the Narrative sound area (a more thorough definition of the sound areas is presented in Chapter 4).

Dykhoff notes:

> The spectators' imagination is by far the best filmmaker if it's given a fair chance to work. The more precise a scene is, the more unlikely it is to affect the audience emotionally. By being explicit the filmmaker reduces the possibilities for interpretation. […] With a minimal amount of visual information and sounds suggesting something, you can get the audiences' imaginations running. (Dykhoff, 2003)

There are many examples of this style of feature film sound design, but a notable example is the sounds associated with the dinosaurs featured in *Jurassic Park* (1993) (Sound Designer and Re-recording Mixer – Gary Rydstrom), which are seen on-screen for only 15 of the movie's total 127 minutes – a little over 10% of the film's total running time; whilst their mysterious 'off-screen' sound is heard by the audience long before they eventually make an appearance (Van Luling, 2014).

Regarding audience emotions being evoked by the soundtrack, Dykhoff goes on to make a highly relevant point:

> It's interesting to speculate about how much information the trigger must contain and how much it actually triggers. (Dykhoff, 2003)

An exploration of the existing literature on emotions and film would seem to suggest that the understanding of the relationship between the overall organization of a soundtrack and the emphasis within the mix – and the resulting emotions evoked in an audience – is still very much in its infancy; even if work on the correlation between emotion categories and types of sounds, or emotions and the acoustic parameters of sounds in music and speech, has begun to be examined more closely:

> Without doubt, there is emotional information in almost any kind of sound received by humans every day: be it the affective state of a person transmitted by means of speech; the emotion intended by a composer while writing a musical piece, or conveyed by a musician while performing it; or the affective state connected to an acoustic event occurring in the environment, in

the soundtrack of a movie, or in a radio play. [...] emotional expressivity in sound is one of the most important methods of human communication. Not only human speech, but also music and ambient sound events carry emotional information.

(Weninger et al., 2013)

Whilst sounds such as speech, music, effects and atmospheres constitute the traditional groupings of sounds within a moving picture soundtrack – especially during its editing and mixing stages – the Four Sound Areas of this research are not intended to be considered as alternative labels for the long-established audio post-production working categories of 'dialogue', 'music' and 'effects' stems.

Rather, they sit alongside instead of replacing those headings; and in any case they do not directly correspond to those categories, by virtue of their being used in a rather different context: the traditional labels of dialogue, music and effects are used primarily in the sub-master 'stems' delivery process before (and after) the final mixing of the soundtrack has been undertaken by the Re-recording Mixer.

As will be seen in subsequent chapters, the Four Sound Areas framework is instead an alternative kind of structure: one that can guide Sound Designers on how best to group emotionally complementary sounds together at the track-laying stage of a moving picture project (i.e. a 'bottom-up' approach); and then help Re-recording Mixers to understand which elements of a mix require emphasis, to increase their ability to enhance, steer or evoke an audience towards a particular area of emotion (i.e. a 'top-down' approach).

2.5 Practical exercise – deconstructing a scene from *Minority Report* (2002) (DVD) using the Four Sound Areas

Director Steven Spielberg's 2002 film is set in the year 2054 and is based on a 1956 short story by the Science Fiction writer Philip K. Dick. The plot for *Minority Report* centres around the experimental 'PreCrime Department', located in Washington, D.C., and the Department's ability to prevent murder through policing advanced warnings of murderous intent in the city's citizens. This information is provided by three highly-developed siblings known as the 'PreCogs', who are kept in a state of suspended animation, floating in a tank of liquid that provides both nutrients and conductivity for the images from their brains to be projected and recorded.

The plot unfolds when the PreCogs visualize the head of the Department, Chief John Anderton, committing a murder. Soon on the run from his own colleagues in PreCrime and seeking to prove his innocence, Anderton discovers the existence of so-called Minority Reports; situations where the PreCog 'pre-visions' are in fact fallible, shown by a difference in their collective presentation of images and characters in the future criminal event. Kept secret to ensure that the experimental PreCrime Department gains nationwide acceptance, Anderton must reveal the truth of this fallibility in the PreCogs to prove his innocence; and also, to prevent any future miscarriages of justice.

The film is sound designed and mixed by Gary Rydstrom (ably assisted by Andy Nelson as his Re-recording partner) and opens with a busy layering of sounds to complement the fast-paced picture editing.

This first scene has examples of Narrative sounds in the dialogue and communication noises between the PreCrime Police officers and the judicial 'Remote Witnesses', as the replayed PreCog visions are examined; and examples of sounds in both the Narrative and Abstract sound areas provide the sound effects of operating the futuristic projector. The associated sounds are of scrubbing backwards and forwards through the vision time-line, and the distinctive room tones, spot effects and atmospheres between the portrayed locations; and there is an example of the Abstract and Temporal sound areas being used together in the Kubrick-esque use of a classical music score to accompany Chief Anderton operating the projector. But the type and placement of this music may not just be a nod to the futuristic, 'space-age' feel created by Kubrick in his landmark 1968 film *2001: A Space Odyssey* – it also accompanies the images of Chief Anderton as the Conductor of an orchestra, as he manipulates the PreCog images and sound through the movement of his hands and arms (e.g. at 00:04:45, 00:05:55 and 00:06:42).

These sound layers have three distinct areas of origination: they emanate from within the projected image (which would seem to accord with sounds placed in the Narrative sound area); from the dialogue and sound effects as the Police officers operate the viewing equipment (sounds from the Narrative and Abstract sound areas); and from the musical score that punctuates the viewing of the PreCog visions in the PreCrime gallery and 'actuality footage' of the future crime that only we, the audience, are able to see (a wealth of sounds that populate the Narrative, Abstract, Temporal and Spatial areas of this study).

This opening sequence serves as a perfect summary of what Rydstrom and Nelson deliver throughout the rest of the film: a carefully balanced, central dialogue (for the majority of the time, serving as a part of the Narrative sound area) that is unchallenged in its intelligibility by any other sound element; thoughtfully understated, futuristic-yet-familiar, spot effects for technological equipment and processes, panned appropriately across the front and rear sound fields; room atmospheres from interior air conditioning and an external suburban atmosphere-track made up of distant traffic, playing children and birds filling the surround channels (all of which contribute to the soundtrack in both the Narrative and Abstract sound areas), along with the noticeable reverberation added to the dialogue, from lines played out in the PreCogs tank area; a room nicknamed by the PreCrime Police as "The Temple", and characterized by its capacious dimensions and hard-reflecting surfaces (the reverberation on the dialogue contributing to the Spatial sound area).

The opening of Rydstrom's soundtrack is playful with the audience: in the opening scenes, he switches the emotional emphasis back and forth between conceivable serenity and veiled anxiety; from the subtle positivity of the sounds of children playing and birds singing in the outside world, to the seemingly cold and antiseptic world of the PreCogs tank room and its monotonous, brooding atmosphere.

It is in this room that the first, but certainly not the last, example of an induced startle-reflex is demonstrated (and where Rydstrom expertly evokes an emotion in the audience that is rooted in fear): the dynamic range of the music (primarily flexing within the Temporal sound area) having been given full reign to accelerate to full-scale, full energy, diminishes progressively down to almost silence, save for a quiet sustained note from the score's string section – along with occasional, delicate drips of water and the distant hum of plant gear (these latter sounds sitting inconspicuously in the Abstract sound area). But by this, Rydstrom is 'setting-up' an unexpectant audience for an explosion of exhaled breath and speech from Agnes, as she suddenly emerges from the water of the Precogs tank (these sounds sitting within the Narrative and Abstract sound areas). In an instant, the maximum dynamic range of the soundtrack is engaged to trigger the sudden, 'heart-stopping' moment in the audience. (DVD, commencing at 00:24:00, with the 'audio shock' at 00:27:25.)

This effect requires careful preparation of the audience; and by utilizing a descent to near-silence just before the metaphorical *coup de grâce* is delivered, Rydstrom has effectively re-aligned the listener's hearing-threshold to a point well below the median soundtrack level. When the sudden, climactic burst of sound is delivered, it is with the maximum dynamic range available to the replay system, but to ears already responding to much lower sound pressure levels.

Given sufficient time with unfamiliar, low-level sounds, and with hearing so highly sensitized, the listening-viewer attempts to make sense of the sounds that they are discerning, but not necessarily recognizing (the sounds are of the Abstract sound area); and a heightened awareness is induced: in essence, the audience is *alert* to danger, and their associated chemical and neural responses are automatically engaged. In short, the audience has been primed to be emotionally induced into fear.

The manipulation of the Narrative, Abstract and Temporal Sound areas to achieve the classic cinematic 'audio-shock', is characterized by the way in which skilled practitioners (e.g. Sound Editors and Re-recording Mixers) use a gentle 'rise-time' to progressively increase, develop and hold their audience in a state of heightened awareness of some impending danger, just prior to the scene's denouement; usually through *almost* indistinct Narrative Sound, ambiguous Abstract Sound and high Temporal Sound elements (a product of the *nature* of the sounds, and most importantly, the relative *balance* between the Narrative, Abstract and Temporal sound areas). This condition is held just long enough for unfamiliar sounds to become recognizable, familiar audio 'bearings' to be re-established, and those markers that suggest to the listening-viewer that there is no imminent danger, to be reinstated. During which time, the unconscious state-of-readiness within the audience – the induced 'fight or flight' instinct – subsides, returning to a near normal level; only for a fast ramp of the soundtrack from (usually) the Narrative sound area to unexpectedly communicate the sudden reappearance of mortal danger.

Minority Report is an excellent example of such skilful sound-blending; with the classifications used for the Four Sound Areas framework readily identifiable within Rydstrom's soundtrack. He achieves several points of predetermined emotional impact, aided and abetted by his clean, precise and uncluttered sound design and the thoughtful, well-balanced and smooth mixing of the movie's audio overall. Such clarity is technically impressive, given the busy nature of the soundtrack at key sections of the film.

2.5.1 Questions

- What emotions were evoked in you by the sound design of the opening sequences of *Minority Report*?
- What influence do you think that had on the rest of the movie?

2.5.2 Discussion points

- What is the fundamental difference between affect and emotion?
- How do sound effects differ from music in evoking emotion?

Notes

1 A traditional and frequently heard idiom amongst film industry technicians wanting to highlight the importance of the soundtrack is 'No one ever came out of a cinema whistling a two-shot'.
2 Director Alan Crosland and Sound Engineer Nathan Levinson had completed a movie for Warner Brothers a year earlier – *Don Juan* (1926) – that used the same Vitaphone sound playback system as *The Jazz Singer* (1927). However, although the soundtrack of *Don Juan* was synchronized to picture, it consisted solely of music with no speech from the actors.
3 There is a famous Hollywood story that suggests the composer Arnold Schoenberg once wrote a film score thinking that a feature film would subsequently be made to match his music.
4 As well as being the Sound Designer and Re-recording Mixer, Murch also picture-edited *The Conversation* (1974) and *Apocalypse Now* (1979). He won an Academy Award for Best Sound Mixing on *Apocalypse Now*.

3   Validating a new approach

Introduction

This chapter outlines the methodology used for validating the Four Sound Areas framework, and this approach may be of particular help to those students who are contemplating, or completing, a PhD by Practice.

It also summarizes the challenges that currently exist for practice-based researchers, lists the main limitations of this study's research, and describes the pressing need for an interdisciplinary approach.

3.1   The methodology for validation

As outlined in Chapter 1, this is a study into emotive sound design within contemporary moving picture audio production and post-production; and one that investigates a new approach to sound design whereby a Sound Designer can consider structuring, and a Re-recording Mixer can consider emphasizing, specific elements of a moving picture soundtrack to steer and intensify the intended emotional experience for listening-viewers. At its core, this study investigates the relationship between the creative process of mixing moving picture soundtracks and the intensity of emotions elicited by that final film.

The Four Sound Areas framework originates in the way that through experience I came to create, construct and mix soundtracks. A natural development of this was to investigate whether the framework could also be of benefit to other Sound Designers and Re-recording Mixers, as well as contribute to the academic understanding and analysis of moving picture soundtracks, and aid industry professionals who have a need to interact and communicate their soundtrack ideas with sound creatives.

My methodology started by exploring existing texts from sound theorists and academics, finding references and information for other practitioners' methods of creating moving picture soundtracks, and then identifying similarities and differences with my own approach. I went on to carry out self-critical reflection on my own work, on the basis of this background; all of which aimed to determine both a theoretical and a practical context for the Four Sound Areas framework.

The study subsequently went on to use this framework for the purpose of achieving detailed analyses of three examples from commercial cinema with noteworthy soundtracks.

The chosen films were drawn from a 40-year period of cinema sound development: ranging from the monophonic presentation of *Winter Light* (1963), the ground-breaking surround-sound movie *Apocalypse Now* (1979) and a more contemporary, multi-channel presentation, *Dogville* (2003).

As part of the validation process for the Four Sound Areas framework, detailed descriptions were also prepared explaining how I used the framework to create original soundtracks for three different types of commercial programming: a short film *The Craftsman* (2012) a full length, theatrical release feature film *Here and Now* (2014) and a sports Outside Broadcast, *Commonwealth Games Boxing* (2014); with reflective and critical analysis of the practical use of the framework whilst creating these soundtracks. Appendix 4 also describes using the Four Sound Areas collaboratively on a feature film project, *Finding Fatimah* (2017).

These studies included considering the ease of integrating this new sound design and mixing process into the workflow of existing television production and feature film post-production procedures, i.e. those used by the location sound recording, picture editing, audio post-production and live broadcast transmission stages; and it could be said that the *external feedback* for the creative and operational usefulness of the Four Sound Areas framework came in the form of the acceptance and satisfaction of the commercial clients who commissioned (and subsequently paid) me to sound design and mix these works.

This methodology had a twofold objective: firstly, to show that the Four Sound Areas framework is fit for operational purpose on different styles of commercial programming, from both the point of view of the practitioner and that of the client; and secondly, to provide scope for a detailed examination of a paradigm designed for both practical and theoretical usefulness.

3.2 The current debates around practice-based research

It is now well over 30 years since the first PhD by Creative Practice was introduced, initially delivered in Australia in the mid-1980s[1] and followed shortly afterwards by universities in the United Kingdom. But it is fair to say that during its lifetime, this type of degree has not had an easy passage – with difficulties being felt by candidate and academy alike. Jonathan Carr writes from experience of his own PhD by Practice, noting in his thesis concerned with filmmaking, that he was conscious of:

> persistent questions over the quality of practice-based work.
>
> (Carr, 2015, p. 26)

He also notes:

> While this scepticism certainly shows a lack of insight into the sometimes gruelling, time-consuming and stressful practice of filmmaking, it also hints at the deeper truth: that there is a still a widely held suspicion that practice-based study does not measure up to its more traditional equivalent in terms of academic legitimacy. (Ibid., p. 25)

Even wondering if:

> there is also a degree of 'knowledge snobbery'.
>
> (Ibid., p. 27, quoting Willems, 2010, p. 9)

There are several notable works that specifically examine the concept of *Practice as Research* (PaR) and the challenge such work has had in gaining academic acceptance in some quarters, particularly in the field of Humanities and the Arts (e.g. Brown and Sorenson, 2009, Candy, 2006; Nelson, 2013) the kernel of which would seem to be the need to overcome a lingering institutional distrust of the results such Practice-based research actually presents.

Yet Smith and Dean strike a more optimistic note:

> The turn to creative practice is one of the most exciting and revolutionary developments to occur in the university within the last two decades and is currently accelerating in influence. It is bringing with it dynamic new ways of thinking about research and new methodologies for conducting it, a raised awareness of the different kinds of knowledge that creative practice can convey and an illuminating body of information about the creative process. (Smith and Dean, 2009, p. 1)

The concept of Practice as Research also raises the question of whether the research itself is practice-based or practice-led; and Candy provides definitions for both:

> Practice-based Research is an original investigation undertaken in order to gain new knowledge partly by means of practice and the outcomes of that practice. Claims of originality and contribution to knowledge may be demonstrated through creative outcomes which may include artefacts such as images, music, designs, models, digital media, or other outcomes such as performances and exhibitions. Whilst the significance and context of the claims are described in words, a full understanding can only be obtained with direct reference to those outcomes. (Candy, 2006, p. 3)

She adds:

> Practice-led Research is concerned with the nature of practice and leads to new knowledge that has operational significance for that practice. The main focus of the research is to advance knowledge about practice, or to advance knowledge within practice. In a doctoral thesis, the results of practice-led research may be fully described in text form without the inclusion of a creative outcome. (Ibid.)

But Smith and Dean find Candy's binary-state rather too prescriptive, instead stating:

We do not see practice-led research and research-led practice as separate processes, but as interwoven in an iterative cyclic web. (Smith and Dean, 2009, p. 2)

And then there is a further sub-division to consider, which is that of Conceptual research. Smith and Dean continue:

Conceptual research is more to do with argument, analysis, and the application of theoretical ideas, and is central to humanities research. [...] Practice-led research practitioners who are particularly concerned with the relationship between theory and practice will see this kind of research as being most relevant to them. (Ibid.)

The current state of affairs regarding Practice as Research would seem to suggest that practitioner and academy are likely to remain uncomfortable bedfellows for some time to come; superficially sitting alongside each other, but with a constant, below-the-surface rumbling of subducting tectonic plates, as it were.

But like any argument, there are two sides to consider; and in this instance, both have merit. Nelson comments:

Given the historical divide between theory and practice in the Western intellectual tradition, moreover, it is not surprising that misunderstandings within and without the academy arose when it appeared that arts practices were suddenly becoming acceptable in the research domain. It did not help that, misunderstanding PaR in believing their professional practice self-evidently constituted research, some would-be practitioner-researchers were reluctant to do anything other than they did as established professionals. (Nelson, 2013, p. 25)

It is undeniable that I arrived at my PhD by Practice from an experienced, Practitioner-based point of view; and as someone who was fortunate in amassing a significant number of credits and a sizeable back-catalogue of creative work to my name. However, to the best of my ability, and held to account by a demanding Supervisor, my research was approached with exactly the same rigour, honesty and principles as that of a more orthodox doctorate.

Carr succinctly encapsulates a balance that needs – and should be possible – to strike:

[...] Willems considers that 'if the purpose of research is indeed to create "new knowledge", then that new knowledge has to be "authentic" new knowledge, which is, in my experience, created through the reconciliation of the "academic theory" and the "professional reality"'.

(Carr, 2015, p. 30; quoting Willems, 2010, p. 20)

At its core, my research was (and this book is) very much concerned with the relationship between theory and practice, aimed at academics and practitioners, and desired an outcome that was of benefit to both.

In aligning with Smith and Dean's sentiments, this study is at times practice-*based*, but then at others it is practice-*led*; and yet it is also underpinned by the conceptual nature of the Four Sound Areas proposition.

3.3 An interdisciplinary methodology

Not only does practice-based research need to talk to and satisfy both the academy and practitioners; often there are additional stakeholders – such as commercial clients, audiences and other professionals – who can quite reasonably judge the successfulness of the practice presented.

3.4 Limitations

The research that I undertook for my PhD, as with any work restricted in time and resources, had a number of limitations, which were at least in part dictated by the type of PhD that it was (I wanted to demonstrate relevance for practitioners as well as for academics, and therefore it was based on creative practice rather than, for example, pure theory); and then, after feeling overwhelmed by the breadth of preliminary reading I'd undertaken of interesting works related to my intended topic, the need to maintain focus on a single pertinent question – and avoid 'mission creep' into a vortex of other fascinating themes – became clear.

Therefore, the format of Chapter 2 will be familiar to academics as a literature review by any other name; but whilst it touches upon several large research areas (emotions, voice and emotions, music and emotions) it does so only to the extent to which those areas are relevant to the sharp focus of my research.

Additionally, the number of film examples analysed and presented had to be restricted. Cinema – and television – are global media; available, consumed and enjoyed by almost every race and culture on earth. However, the three major feature film soundtracks chosen for analysis in this study, created by other practitioners, might at first glance seem somewhat mainstream and rather obviously Western in their origin: from Sweden I offer *Winter Light* for analysis, from the United States of America *Apocalypse Now* and from Denmark, *Dogville*.

It is of course impossible to be anything approaching definitive whilst limiting the choice of other Sound Designers' work to just three examples; but the rationale of the selection was thus: *Winter Light* is a black and white art house film, and included because it is not presented in my native language, but in its original Swedish – the dialogue can therefore be considered as an integral function of the soundtrack as a whole, rather than the listener inferring emotion directly from the meaning of the words used by the actors (not withstanding any emotion suggested by tonality, tempo or emphasis; the emotional impact of dialogue and delivery being covered elsewhere in this book). Also, as a monophonic soundtrack, it offers the most basic form of presentation.

In comparison, *Apocalypse Now* is one of the first Hollywood examples of dense, multi-channel sound design and mixing and serves as a direct contrast to the intentionally stripped-back, and in comparison, bare soundtrack of *Winter Light*.

The final choice of *Dogville* is principally down to the understated, clever complexity of its 5.1 soundtrack, and the fact that it is specifically designed to work without the aid of supporting visual cues throughout the movie; an unusual thing to aim for, and difficult to successfully sustain. It is also included to represent the Independent film genre.

Whilst these three films are included here to represent three distinct – and subtly different – areas of film production: art house, Hollywood major and Independent studio, there is also a professional acknowledgement, admiration and respect for these specific pieces of work, and particularly for the protagonists behind them: Bergman (*Winter Light*) and von Trier (*Dogville*) being Directors who conspicuously have sound as an integral part of their creative filmmaking process; whilst Murch (*Apocalypse Now*) is a Sound Designer and Mixer whose work as a Practitioner has done so much to pave the way for sound practice to command attention from the academy.

There are seemingly countless other notable films that could have taken the place of these three examples, similarly, included for their qualities of sound design and mixing, and ready analysis. The inclusion of world cinema examples would have given a much greater sense of latitude; but even close to home, considering the prominence of actuality sound used in French cinema, the routine and total replacement of on-screen actors voices in Italian productions, or the use of sound effects in dark, Spanish fantasy films would undoubtedly have been interesting to explore.

Further afield, the use of non-diegetic sound in colourful Indian 'Bollywood' productions, the exaggerated sound effects in epic Chinese kung-fu movies or the religious and ethnic differences that need to be accommodated by the Nigerian 'Nollywood' cinema industry (a country of some 186 million people and 500 different languages[2]) would equally make for an interesting investigation into the relevance of the Four Sound Areas framework.

Instead, the examples that were chosen for analysis are here because I think they offer a ready opportunity to observe instances of the Four Sound Areas framework in action, within three familiar styles of feature film production, and with distinct contributions from exceptional practitioners.

3.5 Discussion points

The primacy of pictures over sound in motion picture production might arguably be because sight overwhelms hearing from the moment of our birth. As Sound Designer and picture editor Walter Murch wrote:

> We gestate in Sound and are born into Sight. Cinema gestated in Sight and was born into Sound. (Murch, introduction to Chion, 1994)

- To what extent do you think that this subconscious process is a factor in sound often being overlooked as an equal and essential partner in filmmaking?
- When would you consider drawing a listening-viewer's attention consciously away from the pictures, by using sound?
- When would you consider the use of more subtle, inconspicuous sound techniques to be a more appropriate way to convey an emotion, a sense of place, or a mood?

3.6 Practical exercises

- Listen to, and list, examples of moving picture sound design in your favourite films that demonstrate both 'obvious' and 'subtle' sound design in action.
- Describe why the sound design was effective in making you feel the way that you did when you were watching the film you chose.

Notes

1 Practice-based PhDs began in Australia in 1984, when the University of Wollongong and the University of Technology, Sydney (UTS) introduced Doctorates in Creative Writing (Candy, 2006). It is understood that the author's PhD by Practice was the first in the world to specifically address emotions and moving picture Sound Design.
2 The three biggest sectors of the Nigerian cinema industry are those of the Yoruba, Hausa and Ghanaian-English languages (the portmanteau term 'Nollywood' being culturally and geographically more accurately applied to the Yoruba films of Western Nigeria; and the sobriquet 'Kannywood' applied to Hausa films emanating mainly from Kano, in the North of the country).

Part 2
Deconstructing the soundtrack

In the same way that painting, or looking at paintings, makes you see the world in a different way, listening to interestingly arranged sounds makes you hear differently.

Walter Murch.

4   The Four Sound Areas

Introduction

This chapter presents the Four Sound Areas framework in more detail and discusses it in relation to other existing sound theories and categorizations. It then uses it as a tool for analysing the emotional sound design of three major cinematic works: *Winter Light* (1963), *Apocalypse Now* (1979) and *Dogville* (2003).

4.1 Concept

The concept of the Four Sound Areas is the culmination of around 20 years' worth of my thinking about the perceived meaning of 'sounds' when listening to the component parts of a full soundtrack, i.e. the individual mix stems; a thought process that originated from hearing the isolated parts of a full mix as they were being delivered for syndication purposes: a case of watching the same, familiar pictures of a particular project, accompanied only by the sub-mix stems required by that project's 'Deliverables' document. Typically this would include a 'dialogue only' mixed stem (known as 'clean dialogue'), a 'sound effects and atmospheres' only mixed stem (referred to as 'clean effects'), then a 'music with sound effects and atmospheres' mixed stem (the 'music and effects' or 'M&E' mix), and finally a 'music only' mixed stem (called 'clean music').

Until relatively recently, due to a combination of hardware constraints and computer processing power, it was not possible to easily and simultaneously 'buss' and record these mix-stems in one pass on a DAW. Instead, the whole film would need to be watched by the Re-recording Mixer (repeatedly), until all the sub-mixes required for delivery were complete. However, the 'happy accident' of that elongated delivery process – and the consequence of the countless extra hours spent in the mixing theatre – provided the impetus for this investigation.

As a Re-recording Mixer regularly engaged in creating delivery stems, I started to notice how sounds less prominent in the full mix took on a greater significance when either the dialogue, or the music, were not present. That is a logical and obvious statement; but for me at least, there was a beguiling and noticeably different 'feel' to the on-screen pictures. Sometimes, something that revealed itself in the 'component mix' as emotionally interesting or sonically significant, disappeared when it was listened to again, within the full mix.

It is quite usual, certainly when a Sound Designer is looking after all aspects of audio post-production including the final mix, that they will tend to operate to a tried and tested formula: maybe starting with a dialogue edit, back-filling any gaps with room tone and atmospheres, then looking for examples of specific sound effects that need to be added, including footsteps and Foley, and finally – as it usually arrives as the last of the audio assets – fitting the music around the dialogue, making sonic space as necessary by reducing the relative level of the sound effects and atmospheres to accommodate the music; or if necessary, adjusting the frequency response of certain tracks to make space for featured dialogue.

However, I realized that this kind of prescriptive approach that had served me well for many years, didn't take into account the new and exciting possibility of exploring how sounds could collectively work to support a plot, if they were considered not just under a general heading of, say, sound effects or atmospheres, but in their own right as emotional drivers or narrative pointers.

4.2 Definition

My proposal is that the inter-relationship between all of a film's soundtrack elements is contained within four distinct sound areas, which I have called the *Narrative, Abstract, Temporal* and *Spatial,* and this framework is significant on two counts: firstly, it enables a Sound Designer to create and prepare a soundtrack for mixing that is intended to achieve an enhanced emotional impact; (i.e. a 'bottom-up' approach to creating, selecting and assembling the soundtrack elements). Secondly, I further suggest that the mix-balance between the sounds contained within these Four Sound Areas – which is under the total control of the Re-recording Mixer – is fundamental to the success of eliciting and emphasizing a desired emotional response from a listening-viewer; (i.e. in this case, a 'top-down' approach, emphasizing specific sounds from the supplied clips).

To use an architectural analogy, it is as if the Sound Designer is given permission to choose all of the materials that a house will be built from (e.g. sandstone, cinder block or brick) and then they provide the plan for the shape and style of the house (e.g. a bungalow, cottage or mansion) to the Re-recording Mixer.

Meanwhile, the Re-recording Mixer also enjoys creative freedom; and throughout the build of the structure, this can involve them in moderating initial ideas, or pragmatically simplifying the original plan, to ensure that the Sound Designer's vision is successfully delivered.

From analysing the creation of soundtracks using the Four Sound Areas framework – which specifically avoids categorizing sounds in a traditional 'dialogue/ music/effects' manner – the chameleonic nature of sound becomes revealed: i.e. many sounds have the ability to evoke a different emotional effect, depending on their utilization in accordance with their *principle sound area* characteristic; and an appreciation of this last point is fundamental to unlocking the full potential of the soundtrack, when using the Four Sound Areas framework.

For example, in a Bar or Restaurant scene, heard conversation could either be dialogue from the *Narrative* sound area (where it is significant to the plot and storyline, and strongly portrays the emotions of the principle characters); or act as murmured conversation from the *Abstract* sound area in the background (to illustrate the presence of other patrons, and the general atmosphere of the place); be acoustically processed to form part of the *Spatial* sound area, (perhaps appearing to be spilling-in from the next room, or outside of the premises, perhaps inducing curiosity of what is to come or going on elsewhere); or utilized principally for its *Temporal* quality (depending on whether the foreground or background speech in question is fast or slow-paced, it can inform the listening-viewer of the energy level of the scene, indicating whether the energy of the situation is one of calm or of being aroused).

However, as already suggested, the *types* of sound utilized in a soundtrack, and the soundtrack's overall *balance*, cannot exist independently of each other. Put more simply, intrinsically a sound (*its type*) is either loud or soft in the mix (*its balance*): it is either conspicuous, or it is submerged.

Specifically, the successful evoking of emotions through the Four Sound Areas framework is not only due to the choice of sound by the Sound Designer. Evocation is most reliably achieved when the Re-recording Mixer *emphasizes* the specific sound area (i.e. its relative level in the mix) at key moments in the soundtrack's timeline – the extent of which is controlled by the Re-recording Mixer. The Sound Designer's choice of evocative sounds alone cannot complete their emotional task without them also being brought to prominence in the mix by the Re-recording Mixer.

And so, the Four Sound Areas are described in this way:

The Narrative Sound area is concerned with sounds that are used to communicate meaning or insight. Dialogue and commentary are important examples of this area in the sound mix, but it may also include certain diegetic music, as well as symbolic and signalling sounds such as the ringtone of a telephone, ambulance sirens and so on, as their meaning is clearly defined, almost like a language.

A sound is considered to be within the Narrative sound area if it is used by the sound designer to:

i) confer significance to a narrative point or event
ii) conspicuously draw attention to a plot point
iii) describe an action or an event
iv) be meaningful to plot progression, with its significance readily understood by the listening-viewer.

The Abstract sound area is concerned with sounds that are less codified in their meaning, such as atmospheres, backgrounds, room tones and synchronous and non-synchronous sound effects; as well as abstract diegetic music, where the music is being used as an emotive, atmospheric device rather than for signalling something specific. In the case of speech or vocal sounds, these are sounds that are chosen for the emotional effect of their tonality or inflection, rather than being recognizable in a literal or language-related way.

A sound may be considered to be within the Abstract sound area if it is used by the sound designer to set a mood, a characteristic or a theme for a scene, without intending the listener-viewer's attention to be consciously drawn towards its presence in the mix.

The Temporal Sound area is concerned with the temporal evolution of the soundtrack, through rhythm, pace and punctuation; such as the non-diegetic music score or a specific sound design element with a strong rhythm, and it is characterized and contrasted by the difference between high rhythm, fast pace, high structure and, conversely, slow pace, loose structure and low rhythm. Although this can mean that the Temporal sound area can envelop many types of sound effects, music and dialogue, a sound should only be considered as being categorized as part of the Temporal sound area when its temporal evolution, e.g. its intended rhythm or pace, *is its most important contribution to the sound design.*

Dialogue for instance may be fast or slow-paced, for intentional emotional effect. At one end of the temporal axis, high conversational energy can require that the audience remain highly attentive: screenwriter Aaron Sorkin (*A Few Good Men* (1992), *Charlie Wilson's War* (2007), *The Social Network* (2010)) is well-known for his fast-talking characters. Describing the speed of speech in *The Social Network* (Sound Designer and Re-recording Mixer – Ren Klyce), Sorkin comments:

> We started at 100 miles an hour in the middle of a conversation and that makes the audience have to run to catch up [...] We were always running ahead. I'm not writing something that's meant to be read; I'm writing something that's meant to be performed. (Grosz, 2011)

In contrast to this is the dialogue of Hope Dickson Leach's 2016 film *The Levelling* (Sound Designer and Re-recording Mixer – Ben Baird). Film critic Mark Kermode made explicit reference to the tempo of the dialogue on BBC Radio 5's *Kermode and Mayo's Film Review* show:

> It's about people failing to communicate verbally and there are great gaps between what they are saying; and it's in those gaps and it's what is happening in those very loud silences, the silences in which all that stuff [is] going on in the background, the sound of the farm, the sound of the insects. [...] It is a film about the space between words [...] a film that is absolutely built from the soundtrack upwards. (*Kermode and Mayo*, 2017)

And then there is some of the most challenging work for a dialogue editor to manage: overlapping lines. The 'where', 'when' and 'by how much' of the overlaps all contribute to the pacing of the final dialogue track – which is particularly challenging when the picture editor has not been sensitive to the needs of the dialogue editor when cutting the scene.

Regarding a less constrained, impromptu dialogue delivery by actors, film Director Robert Altman's style of utilizing multiple personal radio microphones instead of relying on boom operators for scenes containing many characters, broke new ground in movies as early as *Nashville* (1975) (Supervising Sound Editor William Sawyer, Re-recording Mixer Richard Portman) and also *Gosford Park* (2001) (Sound Designer Nigel Mills, Re-recording Mixer Robin O'Donaghue). Here, it was technology – specifically the use of 24-track audio tape recorders on set (the kind of large machines used at that time in studios recording multi-track music, rather than the more modern, and considerably smaller, multi-track, solid state devices seen on-set now) – that enabled the Director to work in a much freer fashion than his predecessors would have been able to. Actor Clive Owen was one of the cast of *Gosford Park*, and he recalls:

> He [Altman] said: 'Stop. Actor says line, actor says line. Nightmare.' So, he sent Derek Jacobi through the middle of the scene, chasing a dog, or carrying a tray or something, just to mess up the neatness of it all. I've worked with directors who, if you put more than a few people in a scene, they're like: 'How am I gonna do this?' Now look at *Gosford Park*. Twenty people in a room, each with their own thread and it looks like he [Altman] just happened to catch it all. It takes an amazingly smart brain to do that. (Gilbey, 2015)

Certainly, it takes a Director with a smart brain; but equally, one supported by a fastidious Production Sound Mixer (Chris McLaughlin on *Nashville* and Peter Glossop on *Gosford Park*) and an attentive Re-recording Mixer, who decides who the audience actually gets to hear speaking at any particular point in the mix.

The Spatial Sound area is concerned with both the positioning of sounds within the soundfield of film and television programmes, *and* the space placed around the presented sound: from the convention of on-screen, front Centre-focussed mono-phonic dialogue, the front Left–Right stereophonic imaging of events to the left or right of the projected image, through to enveloping atmospheres placed in the sur-round channels, or specific out-of-vision sound effects, spot effects or dialogue, behind, or (in the case of proprietary playback systems such as Dolby Atmos[1] or DTS: X[2]), above the audience. Almost all sounds inherently display some kind of spatial characteristic, but they will only be considered as part of the Spatial sound area when this characteristic is their most important sound design aspect.

As stated earlier, this concept of any sound being considered part of a particu-lar sound area *depending on its primary emotional value to the soundtrack, or its contribution to a deeper understanding of the plot* is fundamental to gaining the most benefit from using the Four Sound Areas framework. So, comparing how a similar sound may be used for effect in the different sound areas, consider the sound of a pack of dogs barking.

As an example of *Narrative* sound, picture a scene in which a prison inmate is seen cutting stealthily through a perimeter wire fence, under the cover of darkness. The sound of dogs suddenly barking immediately suggests to the listening-viewer the danger of the prisoner being discovered escaping; and the Re-recording Mixer is likely to balance the sound of the dogs for prominence in the mix.

Conversely, the sound of dogs excitedly barking in a countryside setting, where in the background we see a local fox hunt assembling (but not serving as the main feature of the scene or shot), would be considered in this instance to be part of the *Abstract* sound area: it is helping to 'set the scene'; and the Re-recording Mixer is unlikely to make this sound prominent in the mix, so long as the hunt is in the background.

But it might be that it is the speed of the dogs barking – the decisive factor of the *Temporal* sound area – that has most relevance for a scene. Imagine a manhunt is seen assembling on screen. When the baying dogs are featured on-screen and the tempo of the barking is fast and furious, it suggests within them a state of high agitation; and an obvious urgency for them to track and seek out their quarry. On the other hand, slower, more punctuated barks might be used in a plains or prairie agricultural scene, where the dogs are more placid, considered and workman-like in their actions of helping Cowboys in the process of rounding up livestock. But in either of these instances – whether it is fast or slow barking – it may be that the Re-recording Mixer chooses to make this Temporal sound the prominent sound – a sonic 'marker' – in the mix.

Finally, a *Spatial* characteristic could be accentuated by the application of stylized reverberation to the sound of the dogs barking, providing a clue to their out-of-shot proximity; e.g. the listening-viewer may infer that the dogs are in an adjoining alleyway, across a valley, inside a vehicle or within a room. This aural suggestion is something that the Re-recording Mixer can create through the use of convolution reverb; and then emphasize appropriately in the mix.

In a nutshell, Narrative and Abstract sound areas may be said to be concerned with the *nature* and *meaning* of sounds. The Temporal sound area is concerned with the soundtrack's *tempo*, measure or meter; whilst the Spatial sound area is concerned with the *positioning* of, and the *perceived space* around, the sounds within the auditioning soundfield.

Again, it is important to note that the Narrative sound area does not merely consist of dialogue – in the same way that the Abstract sound area does not only have to contain sound effects.

Summarizing the semantics of the Four Sound Areas, a sound can only really be in one of two states, irrespective of what sound area it is conceptually designated to be 'in' at any given moment. That is to say, it is either conspicuous in the mix, or it is in the background of the mix; because the purpose of any individual sound is either to be noticeable or not at a given point in the soundtrack, as suggested by the Sound Designer, and then implemented through the mix-balance by the Re-recording Mixer.

So, for example, a piece of music that displays both Abstract and Temporal qualities, or a sound effect that displays both Narrative and Spatial qualities, when mixed for its *emotional purpose*, should be considered as belonging to the sound area that contributes the most to producing the *emotional intent* chosen by the Sound Designer.

Emotion from Narrative sound can be derived from the listening-viewer's empathy with the sentiments expressed by an on-screen character, and the manner, inflection, rhythm and pace with which the words are delivered (e.g. the desperation expressed by actor Ralph Fiennes' performance in his portrayal of Count Almásy, attempting to persuade British soldiers to help him rescue his injured lover Katherine Clifton, in *The English Patient* (Minghella, 1996); or the robotic (yet recognizable in human terms), distressed and plaintive calling of Eve, as her companion robot Wall-E remains gravely unresponsive in the eponymously titled movie (Stanton, 2008) (Sound Designers – Walter Murch and Ben Burtt, respectively).

Whilst the spoken word predominantly presents in the front, Centre playback position, other Narrative sound area sounds, when moved around the soundfield, can be deployed to evoke a significant emotional reaction in response to a symbolic or signalling sound, out of frame or behind the listening-viewer (e.g. a doorknock or door-bell, a gunshot, a siren, a ringing telephone, an animal's roar, a hissing or rattling snake or a snapping twig.)

The visceral nature of some sounds (particularly when forming part of the Narrative sound area) can also bring about extreme reactions: a certain section of Danny Boyle's 2010 film *127 hours* (Sound Designer – Glenn Freemantle) caused the reporting of multiple instances of fainting within audiences worldwide. One on-line critic wrote:

> The cracking sound effects really make you cringe.
>
> (Yahoo, 2011)

While *Daily Mail* reviewer Chris Tookey claimed *127 Hours* is:

> the most harrowing, bone-breaking amputation scene in the history of cinema. (Tookey, 2010)

Emotions from sound in the Abstract sound area can be derived from experiencing abstract and atmospheric environmental sounds of the scene in question, the emotional effect of which may be increased through the use of the whole soundfield; as if the audience is within the same location as the on-screen characters. Obvious examples of mood-setting can be readily heard in the sense of peace and safety suggested by the sound of songbirds, a sense of wariness from the presence of wind effects, or a sense of foreboding from the sound of an approaching thunderstorm.

Emotion from Temporal area sounds will often (but not exclusively) be derived from elements that demonstrate qualities of musicality; and where unambiguous musical scores, cues and motifs do appear, their use is usually as originally derived from musical theatre. Its early adoption to synchronized talking pictures meant that music was often used as:

> a means to track a character both physically and emotionally throughout a narrative. [...] this approach can be both emotive (conveying to the audience

how to feel) and functional (providing a form of narrative enunciation or commentary.) (Whittington, 2009, pp. 40–41)

However, Whittington goes on to suggest that the contemporary use of non-diegetic and diegetic music, or score and source music, has become more refined; with the score music (often forming a part of this framework's Temporal sound area) regularly reduced to leitmotifs and used to support a soundtrack constructed principally around a kernel of dialogue (often within this framework's Narrative sound area) and atmospheres or room tones (often found in this framework's Abstract sound area), which are used to establish a setting or location.

Source music, however, can present to the listening-viewer a sense of era or signify a specific place that is directly relevant to the plot, often by offering less-intrusive pointers to the storyline (Ibid.).

The careful combination of both diegetic and non-diegetic music can also add an extra depth and sophistication to a soundtrack: e.g. when a contemporary non-diegetic score is worked around examples of period diegetic music that are being used to suggest a familiar, yet past, setting.

Taking this melodic complexity a stage further, according to the Four Sound Areas framework, music with lyrics has the capability to demonstrate Narrative, Abstract, Temporal *and* Spatial sound area characteristics: e.g. when the sung lyric is used as part of the Narrative sound area; or the melody providing the contours of the song suggests an Abstract, less codified message; or the time-signature of the music itself is used to determine the Temporal nature of the scene; or the Spatial acoustic setting of the music suggests a distinct sense of place or environment. However, its conceptual 'home' within the Four Sound Areas at any one point will be determined by *its primary purpose in the soundtrack.*

The last of the Four Sound Areas is Spatial. This is by its nature a very versatile element of the soundtrack, in as much as it can be directly applied by the Sound Designer or Re-recording Mixer to support or reinforce the other sound elements contained within the Narrative, Abstract and Temporal sound areas. It may be useful to consider it as the audio equivalent of the *depth of field* of a camera lens.

This may be achieved through the use of artificial reverberation added to a sound, to alter the 'acoustic' setting a scene is played in (the processed signal being the Spatial sound area element, as opposed to the clean, original sound); or the addition of discrete sounds in surround channels to signpost a sense of place in filmed drama: e.g. 'birds and bugs' to suggest outdoors; or specific sound sources such as crowd microphones placed at an Outside Broadcast site, which are used to convey audience and arena size in the context of a sporting or musical event.

It is important to note that the use of the Spatial sound area is not channel-dependent, i.e. limited to the surround channels of a delivery platform such as Dolby Digital, configured for 5.1 or 7.1 playback. Indeed, Spatial sound area content may equally be applied to good effect in a monophonic soundtrack presentation. And so, in this way, the three separate yet inextricably linked Narrative, Abstract and Temporal sound areas may be bound together into a cohesive whole through the considered use of the fourth element, the Spatial sound area. However, we need to remember that a sound is only part of the Spatial area if the *main characteristic* through which emotion is portrayed is its spatial characteristic.

In *Touch of Evil* (1958) (re-edited and re-mixed by Walter Murch to sacked Director Orson Welles' original mix-notes in 1998), the Spatial sound area conspicuously takes over from the Narrative sound area in one particular scene that is significant to the plot.

Sheriff Hank Quinlan (played by Welles) is attempting to frame Vargas, an innocent bystander on honeymoon with his bride in Mexico, who has been drawn into Quinlan's investigation into a terrorist bomb explosion. Faced with a false drugs and murder charge, Vargas, with the help of Quinlan's long-suffering Deputy, Menzies, draws out Quinlan's motives at an outdoors meeting, at which Menzies is wearing a concealed microphone. Vargas stays out of sight but in range of the radio transmitter, listening-in through a portable receiver with an integral loudspeaker, to Quinlan's incriminating motives being transmitted by the hidden microphone. (We need to suspend disbelief at Vargas not wearing headphones.)

As Menzies and Quinlan walk over a bridge, Vargas is forced to wade through water underneath the bridge, and the sound of Quinlan's own voice, heard from the loudspeaker underneath the bridge, is slightly delayed and it echoes, alerting Quinlan that he has been set up. The pre-bridge dialogue remained firmly in the *Narrative* sound area; but at a crucial point, the dialogue switched to the *Spatial* sound area.

Finally, it would be remiss to not mention the *absence of sound* as a sound design and mixing tool. Certainly, as far as motion picture sound is concerned, silence is the beginning (and end) of all audible sound within a soundtrack; and therefore, it is the quiescent state of all of the Four Sound Areas.[3]

Total silence (as opposed to an extremely low effects track that is barely audible), is rarely used in television soundtracks (indeed, in broadcast television, the presence of total silence in a submitted programme can be a trigger for failure of the pre-transmission Quality Control process, requiring dispensation) and is not the focus of this research. It is, however, a powerfully evocative option for the cinema Re-recording Mixer.

The use of silence in the film *The English Patient* (1996) by the film's Sound Designer and Re-recording Mixer Walter Murch is described in Michael Ondaatje's book *The Conversations*. He relates how silence is used in the memorable interrogation scene:

> When Caravaggio says, "Don't cut me," the German pauses for a second, a flicker of disgust on his face. [...] We see the look on the German. And now we know he has to do what he was previously just thinking about. To emphasize this, Murch, at that very moment, pulls all the sound out of the scene, so there is complete silence. And we, even if we don't realize it as we sit in the theatre, are shocked and the reason is that quietness. (Ondaatje 2004, introduction p. xx)

The shock of silence within modern, fully filled sound tracks can be as striking as any sound effect; and even in older, monophonic soundtracks, such as Bergman's *Winter Light* (1963) the fact that silence is so unnatural in real life grabs and focuses the audience's attention on the screen. A similar effect can be achieved with a single effect being surrounded by silence, e.g. the repeated popping of a photographer's flash bulb at the end of *Rear Window* (1954) (Dir. A. Hitchcock/ Re-recording Mixer Loren Ryder); or the use of a 'near-silence' sequence, left hanging in the air during a dramatic moment, e.g. in *Mission: Impossible* (1996) (Dir. B. De Palma/Sound Designer and Re-recording Mixer Gary Rydstrom) Tom Cruise's character Ethan Hunt is lowered head-first into a high-tech vault that has an alarm triggered by any sound that rises above a certain threshold[4] (Wright 2008).

Considering it is capable of such powerful effect, moments of complete cinematic silence are few and far between, but a notable example of absolute silence occurs in Martin Scorsese's *The Aviator* (2004) (Re-recording Mixer Tom Fleischman):

> After Hughes (Leonardo DiCaprio) has locked himself in his office and he begins to hallucinate, Scorsese pulls [requests the Re-recording mixer to fade out] the sound of the scene. Hughes is naked, watching his own films on a loop. We stare at Hughes in a medium-long shot and as he sits in his chair the soundtrack goes mute. Not even the sound of the projector. Not even the sound of Hughes breathing. Nothing. It's a stark moment because at this point, he has sunk so low into depression and sickness that he [is] finally alone. The silence lasts for only a few seconds, but its presence is hard to ignore. (Ibid.)

Whilst its occurrences might be scarce, the total absence of sound at key moments in a film is an effective tool for the Re-recording Mixer to create or emphasize intrigue, engaging an audience, and focussing their attention to the events unfolding on-screen.

4.3 The alignment of the Four Sound Areas with existing sound design theories

The concept and practicality of breaking down a soundtrack into the four designated sound areas – Narrative, Abstract, Temporal and Spatial – can offer Sound Designers and Re-recording Mixers a useful and novel framework to work with; one that avoids considering the evoking of audience emotions solely by sounds traditionally categorized under a general heading of 'sound effects': e.g. the frequently heard use of thunder to signal a scary scene, or the use of twittering birds to denote a peaceful scene.

Instead, this framework allows for the consideration that it is not just specific sounds, but the whole of the audio mix and its level balance that contributes towards the evoking of audience emotions; and that suitably arranging these elements is something that Sound Designers and Re-recording Mixers are in full control of at the track laying and mixing stages; the point at which they add their creative, and emotional, interpretation.

In this sense, whilst still ordered, it is intended that the Four Sound Areas framework should be a freer, less constrained way of considering sound elements than the traditional audio post-production categories of 'dialogue', 'music' and 'effects' stems. The intention being to allow Sound Designers and Re-recording Mixers to more easily 'make the soundtrack of the world and the voice of the living heard.' (Serres, 2012).

4.4 The rise of emotionally expressive Sound Design

Even after the introduction of the first stereophonic, magnetic soundtracks in the 1950s, many US independent theatre chains resisted upgrading their noisy, monophonic optical equipment on cost grounds until well into the late 1960s; even at one stage petitioning major studios such as Warner Bros not to upgrade their audio playback formats (Whittington, 2009).

Indeed, some 30 years after the release of the first ever multi-channel soundtrack, by Disney (*Fantasia*, 1940; Sound Recordist William E. Garity):

> The frequency range and quality of sound in most cinemas was not much better than that of telephones and continued to remain so until the mid-1970s […] until the arrival of Dolby. (Sergi, 2004, p. 14)

It took the introduction of the Dolby Stereo soundtrack in the early 1970s, with its Type-A noise reduction and inherent mono-compatibility, to convince reluctant theatre owners to raise the playback quality within their cinemas to something more akin to the listening experience available through high-fidelity FM music-radio services and stereo record players, that by then, were already in widespread use by the general public. The contemporary cinema-going population had come to expect more from a film soundtrack, given that the advances in both music reproduction equipment and the wider sound production values afforded to popular music such as The Beach Boy's *Pet Sounds* (1966) or The Beatles *Sgt. Pepper's Lonely Hearts Club Band* (1967) albums; both of which carried extensive non-musical sound content, in the form of sound effects designed to enhance the listening experience, resulting from the technology and experimentation that flowed between the two popular media of music and film (Whittington, 2009).

The more subtle use of the surround channel in cinema audio mixing emerged some time after the introduction of Dolby's Surround Sound[5] technology in the late 1970s; at a time when the novelty of science-fiction films with their 'seat-shaker', low-frequency effects (LFE) channel, and the front-to-back swoosh's of spaceships, lasers and rockets – made possible by Dolby Stereo[6] – had begun to wane. Now Sound Designers and Re-recording Mixers were more able to stimulate the audience directly, by routinely delivering more prolonged periods of 'background' sound all around the suitably equipped movie theatres.

Listening-viewers were also now occupying an auditioning environment that was affected by enveloping and conditioning soundscapes, and it gained momentum by becoming an integral part of the soundtrack expectations of big budget feature films, as well as garnering the attention of film sound theorists. Whittington wrote:

> For the most part, traditional sound theory does not respond to the new needs of sound analysis, primarily because it does not envision the complex production capabilities of the modern dubbing stage, or the use of multi-channel sound formats [...]
>
> (Whittington, 2007, p. 8)

A situation that the Four Sound Areas framework seeks to redress.

The widespread adoption of multi-channel recording and playback formats have broadened the canvas for both Sound Designers and Re-recording Mixers; and cinema sound is no longer limited to emanating from behind the screen or constrained by the framing of the picture. The immersive capability of surround sound lends itself well to the placing of atmospheric sounds all around the audience and these enveloping atmospheres can be put to good emotional intent.

As Sonnenschein suggests:

> Environmental sounds like water can produce a feeling of cleansing, awakening, while wind can provoke a sense of lack of direction, perhaps a place or moment not to be trusted. (Sonnenschein, 2001, p. 205)

An example of this is the use of desert wind in Sound Designer and Re-recording Mixer Walter Murch's soundtrack for *The English Patient* (1996):

> The desert is a vast place. When you're there, the feeling it evokes is psychic as well as physical. The problem is that if you record the actual sound that goes with that space, it has nothing to do with the emotion of being there. In fact, it's a very empty, sterile sound. The trick in *The English Patient* was to evoke, with sound, a space that is silent. We did it by adding insect-like sounds that, realistically, would probably not be there ... Also, tiny sounds – as tiny as we could get – of grains of sand rubbing against each other ... We took those tiny things and made a fabric out of them. (Théberge, 2008, p. 54)

Desert wind is often put to work by Sound Designer and Re-recording Mixer Skip Lievsay. His use of subtle wind induces an aching sense of loneliness and vulnerability into the sound design of Joel and Ethan Coen's *No Country For Old Men* (2007) for instance; and the understated and suggestive use of wind in the surround channels can be heard to similar effect in Lievsay's other soundtracks for desolate and dark Coen Brothers films, such as *Fargo* (1996), *A Serious Man* (2009) and *True Grit* (2010).

Sounds from the Abstract sound area that are placed in a surround channel often consist of three major elements of the stochastic, natural soundscape, summarized by Treasure as 'wind, water and birds' (Treasure, 2007).

But not all sounds originally heard in the Abstract sound area will necessarily remain fixed in that domain; in some instances, for emotional effect, an Abstract sound area leitmotif can eventually be resolved to the Narrative sound area; and a skilled practitioner can use this to great effect.

In the Steven Soderbergh film *Erin Brockovich* (2000), Sound Designer and Re-recording Mixer Larry Blake fashioned a leitmotif from an indistinct and distant heavy generator hum, which was present in the soundtrack whenever scenes showed locations and homes close to a power generation plant. Erin Brockovich, a humble legal clerk, has proof that the operator of the plant, the Pacific Gas & Electric company, knew that their power station was leaching the toxic and carcinogenic chemical Hexavalent Chromium $((Cr)(VI))$ into the local water supply, resulting in an elevated number of residents facing debilitating and fatal illnesses. Her struggle is to convince locals who have unwittingly been bought-off by the corporate giant to join with her, to bring about a legal class action. At the precise moment that Erin Brokovich finally gets a local mother, helplessly watching her child die from Chromium poisoning, to understand that the Pacific Gas & Electric Company are culpable for her child's demise, the sound of the power plant – seen in the distance – vents a subtle, human-like sigh before diminishing in level.[7]

In this instance, and in Four Sound Areas terms, the Re-recording Mixer moved a significant component of the soundtrack away from the Abstract sound area (the humming of the power plant's 'voice' was sat 'back' in the mix-up until this point) towards a place in the Narrative sound area, and a suitably prominent level in the mix. Emotionally, the 'sigh' of the power plant's 'voice' is introduced to signal to the audience that at last the Mother understands the true cause of her child's incurable condition. A further assumption is that the intended emotional effect on the listening-viewer is to increase the emotions of sadness and compassion already felt and increase the intensity to evoke emotions associated with disgust and anger.

4.5 The framework in use

Initially, I depicted the relationship between the Four Sound Areas by using a simple 3-dimensional representation (Figure 4.1). Whilst the two sound areas of Narrative and Abstract are shown at opposite ends of the same axis (because both of them carry 'meaning') it is important to note that these two sound areas are not merely antonyms of each other; they are distinctly different in their function, and therefore each warrants its own sound area. Whilst this representation is designed as a straightforward way to help visualize the sound areas, it is not designed to represent an actual, well-defined multidimensional space; rather, it is provided as a way to 'imagine' the Four Sound Areas, and also as a way in the first instance to communicate the framework to others.

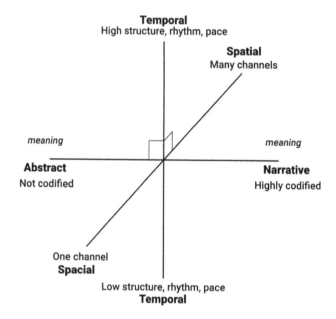

Figure 4.1 A simple 3-axis model representing the Four Sound Areas[8]

The Temporal axis ranges between, at one end, possessing a complex rhythmic structure that is *fast paced* and at the other end a low-complexity, *slow paced* rhythmic structure.

The Spatial axis ranges between describing a sound with a single channel presentation at one end (or a signal with a close, tight presence) and a multiple channel (or a spacious, reverberant signal) presentation at the other.

Although the decision over the *type* of sounds put forward for use in a soundtrack is that of the Sound Designer, their relative *balance* within the sound mix is always set by the Re-recording Mixer (although as noted earlier, increasingly, the Sound Designer and Re-recording Mixer can be one and the same person).

Expressed succinctly by Thomlinson Holman, the job of the Re-recording Mixer is ultimately about

> getting the right sound, in the right place, at the right time. (Holman, 2002, p. 172)

And given what has been discussed in this study regarding the Re-recording Mixer's creative responsibility, it may be appropriate to respectfully add the codicil 'and at the right level'.

4.6 The Four Sound Areas used for soundtrack analysis

As part of describing and verifying the suitability of the Four Sound Areas as a tool for soundtrack analysis, three major feature films, each with notable sound design, were reviewed; and instances identified where the Four Sound Areas might be considered as being present and 'heard' in their soundtrack.

The two feature films used later in this book for self-critique and reflection – *The Craftsman* (2012) and *Here and Now* (2014) – are from the contemporary, independent British film sector; and as a practitioner I have been fortunate to contribute to a wide variety of moving picture projects: ranging from UK and US 'indies', to more mainstream Hollywood studios. But as previously stated, the Four Sound Areas framework is intended to be relevant and applicable to all types of moving picture soundtracks, regardless of oeuvre, genre or budget.

Therefore, the intention in the selection of the following movies for 'pre-existing' soundtrack analysis, was to draw on diverse examples of narrative film-making (including one in a foreign language) and to determine the existence and relevance of the Four Sound Areas in a broader context than any one industry sector, or one particular period of film production, could offer.

During their editing and mixing stages, each of these three films would have been constituted in the traditional way from dialogue, music and effects stems; however, this retrospective analysis of the soundtracks was undertaken to show how this research's Narrative, Abstract, Temporal and Spatial classification of sounds can be used not only for designing and mixing soundtracks, but also for soundtrack analysis in e.g. a part of wider film studies or film criticism frame of reference.

4.7.1 Winter Light *(1963)*

Director Ingmar Bergman filmed *Nattvardsgästerna* (Eng. *Winter Light*) (Sound Designer Evald Andersson/Re-recording Mixer Stig Flodin) between November 1961 and January 1962 and it was released in February 1963. It is the second in a trilogy of films written and directed by the Swedish *auteur* that began with *Through a Glass Darkly* (1961) and ended with *The Silence* (1963) (Figure 4.2).

With a running time of just over 80 minutes, this is one of the shortest of Bergman's feature-length films and helped contribute to a heightened sense of unease that ordinary people were feeling at the time, living in post-war Europe and a world still coming to terms with the cold war, atomic bombs and the break-down of traditional social norms; disconcerting themes that were being explored and exploited in other movies released around this time, such as *On The Beach* (1959) (Dir. Stanley Kramer/Sound Designer Walter Elliott), *La Notte* (1961) (Dir. Michelangelo Antonioni/Sound Designer Claudio Maielli) and *Dr. Strangelove* (1964) (Dir. Stanley Kubrick/Sound Designer John Cox).

Winter Light has a stark and economic soundtrack, monophonic and decep-tively simple in presentation; yet one that at times is bluntly expressive with its

**"BRILLIANTLY DONE...
THOUGHTFUL, ENGROSSING,
SHOCKING FILM...PIERCING,
STARK AND UNSETTLING!
THIS IS A FILM TO SEE
AND PONDER!"**

—*Bosley Crowther, New York Times*

"I ASSURE YOU IT IS A BEAUTIFUL MOVIE!"

--*Brendan Gill, The New Yorker*

**"ONE OF THE MOST MOVING
FILMS I HAVE EVER SEEN.**
Its total effect is shattering!"

—*Philip T. Hartung, Commonweal*

INGMAR BERGMAN'S

**WINTER
LIGHT**

written and directed by INGMAR BERGMAN
starring INGRID THULIN
GUNNAR BJORNSTRAND
MAX von SYDOW
GUNNEL LINDBLOM
A Svensk Filmindustri Production
released by
JANUS FILMS

Figure 4.2 Winter Light, (1963)

use of sound effects and atmospheres, and equally confronting in its use of total silence for dramatic effect. Its black and white pictures reveal its vintage and are as obvious a contrast to the colour of *Apocalypse Now* and *Dogville* as is the disparity is between their richer multi-channel soundtracks and the stripped-bare, single channel source of *Winter Light*.

The film opens with the sound of a single chiming bell, at first calling the faithful to worship and then the church clock chiming the hour of 11 (in the Narrative sound area, firstly signalling a church and then announcing the time of day) at the rural Swedish church of Mittsunda. The turnout is low. Four of the attendees are regulars; the other three have objectives other than ecclesiastical for seeing the middle-aged presiding Pastor, Tomas Ericsson. The camera singles each attendee in turn as the first hymn is sung (predominantly used for its Temporal quality, the music setting the tempo for the Eucharist ritual dialogue that will follow, as well as the pacing of the picture cuts).

Märta Lundberg is in love with Tomas and they have been uncomfortable lovers. Being present at the service means she can be close to Tomas. The other two people with matters on their mind are a young wife, Karin Persson, and her fisherman husband, Jonas.

Tomas is miserably tired, feeling ill with influenza and troubled by a persistent cough. The cough becomes a leitmotif for Tomas's character; it is a Narrative sound area sound that reinforces and serves to remind the audience of his weakened physical and mental state. After the service, when the visiting hunchback Sexton of the nearby village church at Frostnas, Algot Frovik, asks to speak with Tomas regarding an urgent personal problem, Tomas dismisses him abruptly, saying they will maybe discuss the matter later, perhaps after that afternoon's vespers service at Frovik's church in Frostnas.

Tomas receives the Perssons somewhat more civilly when they enter the vestry; Karin, who is pregnant with their fourth child, asking if Tomas would speak with Jonas in private. Jonas is uncommunicative and in a state of silent anxiety and deep depression, the root of which Karin explains is his learning that the atomic bomb is now available to the Communist Chinese; and as he believes they are brought up to hate the free West, for him it is only a matter of time before they deploy their deadly missiles. Karin hopes that as a man of God, Tomas may be able to explain some higher purpose behind all of this, to help Jonas come to terms with what he is feeling. 'We must put our faith in the Lord' ventures Tomas as a platitude, but a scornful look from Jonas ensures that any further conversation remains stilted, with Jonas remaining silent throughout. By the time Tomas tells them 'I understand your anguish, but life must go on ...' the meeting comes to a close, on an inescapable note of foreboding; and Karin asks Jonas to drive her home so that he may return within 30 minutes to talk again, more privately, with Tomas.

The passage of time is an important element of this story (it takes place over a short period of time, possibly only four hours; an equivalent time to that which Christ suffered on the cross, Frovik will later comment) and the physical manifestation of this, represented by the grandfather clock, is a consistent part of the soundtrack in the vestry scenes. At various times, the ticking takes on an abnormal prominence akin to hammering or reduces to the faint hint of mechanization, as if heard from a pocket watch. In doing so it moves between the Abstract sound area (when it sits low in the mix) and the Narrative sound area (when it is exaggerated, suggesting a head-thumping, feverishness in the sickly Tomas) e.g. [**DVD 00:14:45 – 00:15:05 and 00:17:22 – 00:17:52**].

As the Perssons leave, Märta enters, thoughtfully bringing Tomas a flask of hot coffee and sandwiches. Ungratefully, Tomas tells her he already has his own coffee and then they begin to bicker; the central theme being how Tomas is unkind to her whatever Märta does to please him.

In the interval between Märta leaving and Jonas returning, and after spending some time looking at photographs of his dead wife to the prominent accompaniment of the clock (Narrative sound area; symbolizing that time has moved on since her death, even though he has not) Tomas reads a letter, in which Märta desperately offers her love to him as a consolation for his lack of trust in God. The letter is read for the audience to eavesdrop, by Märta, in a disconcertingly prying close-up shot. There are no visual or audible distractions; her voice is rooted firmly in the Narrative sound area as she explains herself. **[DVD 00:26:50 – 00:34:16]**

Jonas is late; so, Tomas pushes the letter aside and falls asleep. When he eventually returns, Tomas wakes to find Jonas standing over him. He has entered not with the sound of a door opening or footsteps on what must be a hard, stone floor, but instead to the loud ticking of the grandfather clock again (Narrative sound area; signalling the importance of the situation: time is literally ticking away for Tomas to help Jonas). The ticking grandfather clock is an important part of the sonic fabric of the vestry: conspicuous when in shot, subdued when not; but when it is emphasized to the point of being overbearing (e.g. when Karin and Jonas Persson first talk with Tomas, or whilst Tomas looks at the photographs of his dead wife before reading Märta's letter) it is only at a high volume level in between, or in the absence of, dialogue **[DVD 00:25:35 – 00:26:50]**

Tomas's first perfunctory, counselling-style questions are clearly unproductive; and so instead Tomas embarks on a discourse about his own crisis of faith. This is clearly of no comfort to Jonas, who leaves without a word. The sound is intentionally taken down to complete silence as Jonas leaves; and the soundtrack not only becomes devoid of clock-ticks, footsteps, hinge squeaks or the vestry door shutting, but it also removes the logical baseline of any interior dialogue scene here: room tone that might contain a faint winter wind blowing through the cracks and gaps in the church doors and windows. This unnatural, complete silence is held for an uncomfortable 30 seconds. **[DVD 00:40:47 – 00:41:17]**

After Jonas has left, Tomas breaks the silence to ask, pompously Christ-like: 'My God … Why have you forsaken me?'; but then the soundtrack returns once more to complete silence, for a further uncomfortable 44 seconds.

Leaving the vestry to re-enter the church, Tomas collapses at the altar, weak from his 'flu; it is his staccato coughing fit that pierces the slow measured pace of the preceding shots. But Märta is lurking, waiting for him. Pleased for the opportunity to cradle, kiss and comfort him in her arms (a visual arrangement piously analogous to a Christian Pietà, a classical artistic theme where a sorrowful Virgin Mary holds the dead body of Christ), Tomas lies limp and unresponsive to her affection as Märta murmurs her love for him. They are disturbed by the sudden entrance of Magdalena Ledfors (the old Widow seen earlier in the congregation) who blurts the tragic news that Jonas Persson has just been found dead by the river. He has taken his life with his own gun, shooting himself in the head.

Tomas immediately assumes his duty as a clergyman, coldly shrugs away his Mistress and pausing only to collect his coat and boots, he leaves for the river, leaving Märta figuratively, and literally, behind him. The sound of his car engine starting as he prepares to make his exit is made prominent in the mix (a significant Narrative sound that tells us the urgency and importance of Tomas's journey, set against the added Abstract area atmospheres and the actual location sound.)

A Policeman is already at the scene and Tomas is asked by the Detective to stay with the body until the infirmary van arrives, whilst he takes Jonas's gun to the Police station. Alone with the body, we sense what Tomas must surely feel: he has been, and continues to be, useless to the situation. Märta arrives and rushes to Tomas, but he sends her away from the covered body. The infirmary van arrives, quickly collects Jonas's dead body and leaves.

Throughout this scene of the suicide on the river bank, which is visually covered in long-shot, the sound is swamped by the unnaturally loud atmosphere of the rushing river (emphasized and forming a part of the Temporal sound area, signalling that the speed of overwhelming events has overtaken everyone) and this maelstrom intentionally drowns out almost every other sound (any possible dialogue, footsteps, clothes rustle, engine noise and door slams are all intentionally submerged beneath the unnaturally loud and disconcerting sound of the fast flowing river). We are eavesdroppers to a difficult situation, albeit from a distance, but it is nonetheless uncomfortable to hear so little of the events unfolding because of the all-pervasive sound of the river's rapids. **[DVD 00:45:04 – 00:49:35]**

After making her own way to the river, then forced to wait in Thomas's car, Märta is driven to her home by Tomas. Although the overwhelming sound of the river is reduced when they enter the car and close the doors, whilst their lips are seen moving, we still cannot hear their voices as they drive off. **[DVD 00:49:05]** At the schoolhouse, Märta prepares cough medicine and aspirins for him and they begin a conversation that quickly develops into a shockingly cruel annihilation by Tomas of Märta's love and intentions towards him. As viewers we wonder what she has done to deserve such cruelty from his lips; and once more we eavesdrop as if embarrassed bystanders. **[DVD 00:52:02 – 01:00:52]**

Märta is emotionally broken by his verbal onslaught, but she continues to describe the unrequited devotion she has for Tomas, which she hoped she had made clear in her letter to him when she said: 'I live for you. Take me and use me. Beneath all my false pride and independent airs, I have only one wish: to be allowed to live for someone else.' Her defence of this hopeless devotion is reminiscent of the lovelorn Helena justifying herself to Demetrius, in Shakespeare's *A Midsummer Night's Dream*:

> The more you beat me, I will fawn on you.
> Use me but as your spaniel – spurn me, strike me,
> Neglect me, lose me. Only give me leave,
> Unworthy as I am, to follow you.
> (Act 2, Scene 1, page 8)

She has been made to endure the cruellest tirade yet from Tomas and the situation has become a mid-winter nightmare for Märta – she realizes that her life with him is untenable, but without him it is unthinkable.

When he is spent, in the quiet aftermath of them both recovering and coming to terms with what has been said, it seems that Tomas has realized he has stepped over a line of common decency, particularly in the way that he has spoken to Märta. He goes to leave – in a pontifical manner declaring that he must be the one to break the news to Karin Persson – but he hesitates at the door, and then asks if Märta will accompany him.

Driving to Frostnas church from the bereaved Persson homestead, they travel without speaking and are forced to wait at a railway level crossing: another moment of significant sound design and visual metaphor. The train arrives and passes close to the car as they wait, the goods trucks visually representing enormous coffins that disappear in the train's smoke and steam as if in a cremation, as do Märta and Tomas; each figuratively lost and alone within the same vehicle. Inside the car, a doubting Tomas, acutely aware of his failure and despair, confides in Märta that it was his parent's dream that he should become a clergyman.

It is the Narrative sound area that is used to convey to the listening-viewer the brutal power of the locomotive engine, causing attention to be directed and focussed on the startling scream of its steam whistle; whilst an Abstract sound area wash – made up of couplings, linkages and chains –provides an appropriate and contextual mechanical underscore.

The Temporal sound area contributes to the scene through the clattering of the railway trucks, their steel wheels on the rails becoming a tumultuous crescendo to the scene; a stark contrast in tempo and volume to Tomas and Märta's subsequent quiet arrival in the village of Frostnas, to the gentle sound of church bells ringing. **[DVD 01:06:15 – 01:07:33]**

In the vestry at Frostnas, as Tomas prepares for vespers that no parishioners have arrived for, with an audience that consists of Blom the organist, Frovik the Sexton and Märta, Frovik suggests cancelling the service. Tomas dismisses this notion out of hand and is determined that order will prevail. Back in the vestry, the crippled Frovik is finally able to reveal his urgent problem. He has read the gospels as Tomas advised him to, as a way to distract his depressed mind from the constant physical pain he must suffer alone; and he tells Tomas that he has come to understand that Christ's suffering was not so much endured on the cross, because that lasted for just four hours. But instead it was his friends disowning and distancing themselves from him, his crying out in despair that his own father, God himself, had forsaken him in his time of need; and he died doubting all that he had preached. It is a clear analogy that suggests Tomas must examine his own conscience and consider how he has conducted himself as a clergyman, a man of God.

Tomas instructs Frovik that the service bell should be rung; and just as the film started with a single, forlorn bell working in both the Narrative and Temporal sound areas, so too the film draws towards its close.

Winter Light is a purposefully bleak film and in helping perpetuate this persistent sense of unease and discomfort, the soundtrack rarely strays from the Narrative sound area: the dialogue, monologue and spot effects all point towards the negativity and despair of the characters and the unresolved and unfulfilled situations they each face. The Abstract sound that appears is often subdued, barely more than room tone fleshing-out the dialogue of interior scenes or low exterior ambience (the level crossing being an obvious exception).

The use of the Spatial sound area is low, too: the church interiors were actually shot in a studio,[9] so the natural reverberation that a microphone would detect when recording in a stone building (e.g. it would be particularly noticeable on voices and footsteps) is absent; and the use of artificial reverberation to compensate for this, e.g. at the opening of the film when we first hear Tomas presiding, is balanced carefully in the mix. There is no attempt to balance for perspective when the sound of Tomas preaching continues over exterior shots of the church **[DVD 00:01:45 – 00:02:10]**

The Temporal sound area is served by several sound sources: at its fastest, the clattering of steel wheels on the train track, at its slowest, the somnolent and doleful church bell; the church organ (there is no score, this is the only music present throughout the film) which influences the tempo of the subsequent dialogue (the dialogue is in the Narrative sound area, the rhythm of which is carefully metered to the point of feeling unusually ponderous at times); the rushing river atmosphere – at odds with the awkward, slow dialogue – is also at a frantic pace; but the metronome of the film is set and re-set by the instances of total silence.

4.7.2 Apocalypse Now *(1979)*

Apocalypse Now is arguably the most significant film for sound design of modern times, with the on-screen credit 'Sound Designer' given to Walter Murch. He was assisted by several notable practitioners, including Re-recording Mixers Richard Beggs and Mark Berger, and an at that time, new-to the-industry rookie (now Director of Sound Design at Skywalker Sound), Randy Thom (Figure 4.3).

Head-of-Department Murch created a soundtrack that would introduce to the film industry the mantle 'Sound Designer', as well the first commercial 70mm Dolby-encoded soundtrack to have split surround channels and an LFE channel.

Figure 4.3 Apocalypse Now, (1979)

By fully exploring the artistic possibilities made possible by Dolby Laboratories' multi-channel audio production system, Dolby Stereo, the audience enjoy a greater involvement through surround sound; and in turn, this helped to pave the way for the widespread adoption of the now ubiquitous Dolby 5.1 standard[10] (Cowie, 1990).

The film broke new ground for sound design in the way that it explored multi-channel mixing. Its complex and dense soundtrack was meticulously planned and executed by Writer and Director Francis Ford Coppola and Murch, who also picture-edited the movie – a dual-role he had previously undertaken on Coppola's films, *The Conversation* (1974), and subsequently reprised by completing both picture and sound editing on *The Godfather Part III* (1990). This double role was later undertaken with Director Anthony Minghella on *The English Patient* (1996), for which Murch achieved the unprecedented accolade of winning Academy Awards for both picture and sound editing (IMDb).

The sound of *Apocalypse Now* moves seamlessly from mono dialogue (contributing to the Narrative sound area of this study), stereo music (contributing at times to both the Abstract and Temporal sound areas of this study), frontally spread L-C-R narration (again, contributing to the Narrative area) and rear-surround ambience and low-frequency effects (contributing at times towards the Narrative, Abstract and Spatial areas).

Throughout the film, we are afforded insight into the emotional condition of the principal character Willard; but examples of other characters include one of Willard's boat crewmen, Lance B. Johnson, during his LSD trip whilst under enemy fire (in Four Sound Area terms, at this point in the soundtrack there appears to be a high Abstract and Temporal sound area emphasis in the mix – Murch perhaps suggesting emotions of isolation, alienation and hopelessness to the audience) and the US Soldier and marksman Roach, whose hearing is so acute that he echo-locates the position of a taunting Vietcong sniper by the sound of their voice alone. Roach's focussing is signified by the fading-out of the surrounding battle sounds (even though we continue to see rocket and tracer fire track across the sky, and movement from actors that previously warranted Foley effects), leaving only the sniper's catcalling. This is rooted in the Narrative sound area – even though we cannot clearly hear what is being shouted, we can understand the tonality and inference of this as a derisive, dangerous sound. By now, Murch has the audience listening to an almost completely Narrative and Spatial soundtrack (the Abstract and Temporal sound areas are removed from the mix for effect), leaving the listening-viewer as focussed in their hearing as Rifleman Roach. **[DVD 03:32 – 04:35]**

Historically, the significance of *Apocalypse Now*'s sound design is that it marked a turning point in the way that American studios produced and delivered cinema soundtracks, due in no small part to the multi-channel possibilities afforded to Sound Designers, through the utilization of Dolby Laboratories' innovative technology:

> In the 1970s and early 1980s, Dolby achieved nothing less than a comprehensive industry-wide transformation, from studio attitudes to sound, filtering through to filmmakers' creative use of sound and audience expectations.
> (Sergi, 2004)

The film opens with the sound of synthesized rotor blades, processed to alter their pitch, tempo and timbre (operating in both the Abstract and Temporal sound areas at first), panned front and rear, and set against images of real helicopters passing through frame in slow-motion (thus transitioning the synthesized rotor 'chops' towards the Narrative sound area, as the helicopters register on screen). The introductory music of a song (further use of sound in the Temporal sound area) is started at a time motivated by shots of Napalm flames devouring jungle and corresponding to a bleak lyric that ironically begins the film with the words: 'This is the end.' **[DVD 00:01:13]** With Temporal sound balanced in this way against Narrative sound, Murch begins to demonstrate the range and extent of *Apocalypse Now*'s soundtrack.

Emotionally, Murch positions the audience from the outset to expect feelings associated with negativity: particularly pity, sadness and worry. Later, as the film progresses and the plot develops, these increase in intensity towards feelings more associated with embarrassment, shame and guilt, ultimately culminating in the evocation of disgust and anger.

Pictures are overlaid and merged from the flames back to rotary blades, but this time we see that they are the air conditioning ceiling fan of Willard's Saigon hotel room. The distinctive chopping sound of these blades (their pace and punctuation being carefully designed Abstract and Temporal sounds, to remind us of Willard being in the jungle) are reintroduced firstly in their synthesized form, and then seamlessly segued into the actuality of a single, real helicopter engine (whose nature therefore becomes that of the Narrative sound area), passing close-by and overhead the hotel room, located in down-town Saigon. This transition acts as the vehicle to bring the audience from experiencing the sound of the psychotic imagination of Willard (with its characteristic of Abstract sounds and the rhythmic nature of the Temporal sounds, emphasizing an emotion of isolation from the here-and-now, accentuated by the absence of Narrative Sound) and into the 'here-and-now'. Real-world atmospheric sounds of the city (contributed by sounds in the Abstract sound area) filter in from outside of the hotel room. Motor vehicles, traffic-cop whistles, pedestrian murmur, insects and birdsong; some of which will segue back to jungle sounds alongside Willard's thoughts:

The car horns as frogs, the scooters as mosquitoes.

(Thom, 2011). **[DVD 03:52]**

The first spoken words heard are Willard's off-screen narration (from the Narrative sound area), and these are presented as being internal; from within his own thoughts. This Narrative sound is set against a quiet backdrop of sounds in the Abstract sound area, with an exterior-to-the-room atmosphere; punctuated by three synchronous, room-interior sound effects: Willard catching a fly (Narrative sound), Willard scorching his wife's photograph with his cigarette tip (Narrative sound) and Willard swallowing a drink (Narrative sound). This emphasis on the Narrative sound area brings about a sharp focus on Willard's words, and the lack of activity in the Temporal sound area acts to uncouple the monologue from the other mix components, removing any sense of distraction or temporal progression. [DVD 00:04:10]

In turn, an emotional shift begins to take place at this point in the soundtrack, initiated by the sharp focus being brought to bear on Willard's private thoughts: the viewer is emotionally led from a position of uninformed compassion to one of involvement and interest.

The opening act ends with the re-introduction of Abstract area sounds that are once more internal to Willard's thoughts; corresponding to his monologue narration and that of a jungle combat-zone he knows all too well. As his narration (part of the Narrative sound area) comes to a temporary conclusion, faster-paced music (utilized by the Temporal sound area) returns. In a frenzy of either drunken or drugged movement (possibly both – the camera has earlier panned across not only a liquor bottle and a glass, but also a spoon that might suggest narcotic use), and at the end of badly co-ordinated Martial Art forms and impulsive choreography, Willard punches the mirror. The smashing of the glass we see and hear (sounds from the Narrative sound area), but his subsequent screams of pain and mental torment, we see but do not hear. The scene ends on music alone (utilizing the Abstract and Temporal sound areas). The emotions Murch presents are related to Willard's anger: his rage, fury and loathing; evoking in the listening-viewer emotions related to sadness: anguish, despair and shame.

[DVD, sequence ends at 00:07:16]

The listening-viewers reference point as to what is 'real' and what is a function of Willard's psychotic imagination is regularly challenged. As the tempo of the score progressively increases to fever pitch (the music working hard within the Temporal sound area), the audience is being conditioned for the confusion and conflicting emotional triggers that are to follow; and as the boundaries between the presentation of external sounds and internal feelings are increasingly blurred, the audience is directed towards emotions such as sadness, fear, anger, surprise and disgust, with perhaps an intention on the part of Murch to evoke within the audience a sense of guilt, set against a background emotion of tension.

Given the Narrative nature of Willard's commentary, the viewer is anchored by the logical thoughts that he clearly and lucidly communicates. Yet, intentionally, this familiar language and sense of order is distorted as a chaotic realm of disorientating and ambiguous Abstract sound area sounds intrude on the calm monologue, with a counterpoint of other readily identifiable Narrative sounds. The conclusion to this intense opening act is in fact only the first climax of what is a carefully arranged manipulation of the viewer's emotional journey, designed masterfully by Murch.

4.7.3 Dogville *(2003)*

The Lars von Trier film *Dogville* (2003) (Sound Designer and Re-recording Mixer Per Streit) is a story of deprivation and exploitation, set in the isolated and Depression-hit 1930s American township from which the film takes its name. Not

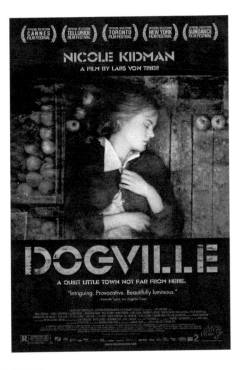

Figure 4.4 Dogville, (2003)

only unusual from Streit's sound design point of view, it is unusual visually. Set out on a large sound stage, all of Dogville town's locations are simply painted in white outline on a dark studio floor, and minimal scenery and props are used to augment and support the actors' performances (Figure 4.4).

The film is driven by a narrator who leads us through the prologue and nine subsequent chapters (his monologue being a function of the Narrative sound area) and initially, it is mixed against a gentle, but cheerful, background of chamber music (used by the Temporal sound area to set a rhythm and pace for the scenes we see, along with gentle Abstract sound area atmospheres). Within the sparse, stark, soundtrack there are little or no visual sources for the spot effects such as door knocks and shop-bells, or latches, hinges and handles, as the actors move around the small town – indeed the actors actually mime opening and closing doors, but nonetheless, we hear these Narrative sound area sound effects and Abstract sound area atmospheres as befitting the on-set actions: such as the United States President's address being listened to on a period radio in the Doctor's house (in Four Sound Area terms, classified as being in the Abstract sound area, as the words of the President are not important to communicate meaning; what is more important is the sense of period that their sound conveys). Footsteps added by the post-production process of Foley (Narrative area sounds) suggest different surfaces underfoot, and after the initial, idyllic scenes, with their Abstract sound area 'backwash' atmospheres, even the time of the day and season of the year are represented by the authentic sounds of birdsong and insects (which all contribute to the Narrative sound area).

As the film's theatrical tableau unfolds, Narrative, Abstract and Temporal sounds are balanced to both support the characters' actions and emphasize the oppressiveness of their progressively enslaving relationship to Grace; the central female character who arrives from out of town unexpectedly, and with a personal story to hide. The Spatial sound area initially remains a constant; predominantly the reverberation that is a function of the natural decay time of the sound stage that the film was shot in.

Steadily, the feeling of comfort and security suggested by the atmosphere 'bed' of light wind and birdsong – which has been the mainstay of the Abstract sound area up to this point – loses its prominence in the mix; and is instead, by degrees, replaced with a mood of heaviness and burden. The strength of the wind rises and deepens in pitch and the usual sentinels of safety – the songbirds – stop singing. A storm is brewing, metaphorically, and the balance and nature of the sounds in the Abstract sound area (which includes the atmospheres), relative to the Narrative sounds (which include the dialogue, Foley and spot effects) and the Temporal sounds (which include the music), reflects this. The Abstract sounds become lower in the mix, and also oppressive in nature.

As Grace's living conditions worsen, the balance between the Narrative, Abstract and Temporal sounds of earlier scenes, demonstrated by the relationship between the light, Abstract sound atmospheres, present in the mix alongside the regular and harmonious Temporal sound interludes – in the form of the string quartet musical score – changes; and as the amount of this Temporal content progressively reduces, an increased awareness of the enveloping chill wind, now more noticeably present in the Spatial area, pervades. Even though overall the Abstract sound area is lower in volume within the mix compared to the level of light wind and birdsong found in earlier scenes, it fills the space created by the reduced use of the cheerful score (utilized by both the Abstract and Temporal sound areas); and the absence of this happy and hopeful music is in direct proportion to the rising unease that pervades the vindictive syntax of the dialogue (which forms part of the Narrative sound area) that is directed towards, and about, Grace. This contrast may be clearly heard by comparing the Abstract and Temporal sound areas of Chapter 4 **[DVD, commencing at 00:54:58]** with those in the film's Chapter 7 **[DVD, commencing at 01:33:10]**

However, perhaps the most shocking event up to this point is witnessing Grace being raped for the first time **[DVD commencing at 01:30:55]** which corresponds to the soundtrack becoming progressively, and eventually, silent. As the camera zooms out to a town-overview wide-shot, we see her violation being perpetrated whilst the town remains symbolically unhearing and steadfastly uncaring to her plight; Sound Designer Streit powerfully suggesting the emotions of isolation, despair and hopelessness to the listening-viewer.

In *Dogville* it is not only the Narrative sound area that is the primary pointer to the intended emotions of Streit's soundtrack, or the setting of its moral compass. It also actively involves the Abstract sound area. As the audience witnesses a series of increasingly distasteful events, audio reference points are designed to evoke emotions of sadness, fear, anger, surprise and disgust in the listening-viewer, as well as embarrassment and guilt, and malaise and tension.

Wind and water (in the form of rain), birds and bugs (all within the Abstract sound area) are used extensively in the surround channels (containing Spatial sound area sounds) to illustrate the story's passage from initially sunlit, untroubled times (signified by the sounds of summer birdsong and insects) towards the dark and sinister turn-of-events that track Grace's enslavement and her ritual abuse by the citizens she works for. This is illustrated particularly through the light, summer breeze that progressively becomes a chill, howling wind that accompanies the prominent Narrative sound area; and is noticeable over time in the surround channel, as the purposeful distraction of the earlier Temporal area score diminishes. And as if to set the tone of the mood, each chapter point – shown by a fade-up from black to a simple caption-card – is accompanied by an appropriate atmosphere that is present in the surround channels first, ahead of the scene we are about to see. (In this study's terms, the Abstract sounds here are actually contributing to the Spatial sound area, whilst remaining Abstract by nature.)

Both the monologue commentary and the actors' dialogue (both of which form a part of the Narrative sound area) are recorded in close proximity (the narrator in an acoustically treated studio, speaking almost certainly into a large diaphragm cardioid microphone; the actors' voices recorded on set through concealed personal lavalier microphones and on cardioid or hyper-cardioid boom microphones) and few instances of realistic audio perspective are used on speech; instead, a sense of distance and isolation is conveyed by the, at first, ambiguous Abstract sound leitmotif of distant pile-driving, which as it becomes more apparent in the mix, and metronomic in its presentation, crosses-over to become part of the Temporal sound area. It is some considerable way into the film, well after the sound has registered with the listening-viewer, that we are told that it is the sound from the construction site of a new State Penitentiary some miles away. This sound therefore serves as a constant, and increasingly insistent, reminder to both the Dogville town-folk, and the audience, that justice is figuratively still present, albeit at a distance.

Crucially, along with *Apocalypse Now*, *Dogville* demonstrates that it is not only the *nature* of the sounds used in the Narrative, Abstract, Spatial and Temporal sound areas, but their *balance* within the mix (i.e. their relative levels) that determines the evocation of certain emotions within the listening-viewer.

Operationally, the mixing of the sound track for *Winter Light* is noticeably less subtle in comparison to a generation-later *Apocalypse Now* or the two generations-on of *Dogville*; and this is hardly surprising: tastes and styles change, improvements in equipment have granted a greater expressive freedom; and as an art-form, cinema has always been subject to, and benefitted from, progressive and empirical change.

But all three of these films confront our sensibilities as an audience – they demand and expect our full attention; and all three have conspicuously utilized sound as a vehicle for not only presenting their own verisimilitude, but to also engage and involve the audience through their emotions.

4.8 Discussion points

- Considering *Winter Light, Apocalypse Now* and *Dogville*, which, if any, of the four sound areas benefitted from being within a multi-channel soundtrack, compared to being in mono?
- Which, if any, of the four sound areas remained unaffected?
- What, if any, benefits to the telling of their stories did the multi-channel presentations have over the mono soundtrack? Why?
- In *Dogville*, how did the atmospheres, sound effects and Foley work to solidify the physical space around the actors; and how effective was it?

4.9 Practical exercises

- Listen and describe the way in which the mono soundtracks for *The French Lieutenant's Woman* (1981) and *The Shining* (1980) are used for emotional intent. Comparing them to *Apocalypse Now*, how might a multi-channel soundtrack have altered the viewing experience of them both?

Notes

1 Dolby Atmos was launched in 2012 and adds height to the surround sound capabilities (the Spatial sound area) of a soundtrack, through the use of dedicated ceiling speaker feeds. It represents an opportunity for sound designers to create a true 3D soundfield, with sound 'objects' mapped to move ('pan') around the available theatre speaker array, although ambient sounds and dialogue that do not move dynamically in the soundfield are still mixed conventionally. Professional Atmos systems in mixing theatres and movie theatres support up to 128 sound channels and 64 speaker feeds. Blu-Ray soundtracks require a separate mix from Theatrical master soundtracks for Dolby Atmos Home Theatre set-ups, which currently support a maximum of 24.1.10 (surround/LFE/ceiling) channels. For home use, a spatially coded sub-stream is added to Dolby Digital Plus or Dolby True-HD soundtracks.
2 Dolby Laboratories' commercial rivals DTS launched its own version of height-enhanced surround sound called DTS: X, in 2015, which promised a distinct advantage for post-production budgets: DTS master soundtracks are compatible with all down-mixing requirements, removing the need for costly re-mixing for home theatre formats. Whilst DTS Home Theatre supports 11.2.4, most home receiver manufacturers are standardizing on the 7.2.4 configuration to prevent a costly 'format war'.
3 Respectfully acknowledging both the philosophical standpoint of Parmenides 'Nothing comes from Nothing' (*ex nihilo nihil fit*) or the quantum physics theory regarding 'nothing' as being an unstable quantum vacuum containing no particles.
4 The vault scene in *Mission: Impossible* (1996) surely owes a great deal to the near-silent scene of a much earlier film: *Rififi* (1955) (Dir. Jules Dassin/Re-recording Mixer Jacques Lebreton), a French 'Heist' film that puts a cleverly constructed soundtrack to use during the near-silent 'breaking-in' sequence.

5   'Split surround' (a term used by Dolby to distinguish it from 'surround sound' that had been part of the sound mixing vocabulary since Disney's *Fantasia* (1940)) was first tested on the movie *Superman* (1978) (Supervising Sound Editor Chris Greenham, Re-recording Mixer Gordon McCallum); and *Apocalypse Now* (1979) was one of the first formal releases of the modern era with three channels in the front (Left, Centre, Right) and two in the rear, Left surround, Right surround). There were typically five speakers behind the screens of 70mm-capable cinemas, but only the Left, Centre and Right were used full-frequency, while Centre-Left and Centre-Right were only used for bass-frequencies.

Earlier, *The Robe* (1953) (sound designed by the pioneer Roger Heman Sr.) had a 4-track magnetic soundtrack configured for Left, Centre, Right, Surround playback (Murch, 2020).

6   The term Dolby Stereo was first used for the 1976 release of *A Star is Born*. The four-channel magnetic audio track occupied the space on 35mm film stock previously utilized by the single monaural optical soundtrack. It carried a front Left/Right stereo channel and a front monaural Centre channel for speakers behind the screen and a monaural Surround channel which was used to feed speakers at the sides and rear of movie theatres. Significantly, Dolby Stereo also utilized the Dolby A-type noise reduction system that enabled sound designers to contemplate the use of low-level sound for the first time.

7   In 2010, Erin Brockovich returned to Hinkley, California to investigate claims that the leaking Chromium plume was spreading, despite Pacific Gas & Electric's clean-up efforts. PG&E continues to provide bottled water for the Hinkley residents, as well as offering to buy their homes (Kahn, 2010).

8   It may also be of interest to consider the extent of the correlation and relevance to the broader scope of the Congruence-Associationist Model proposed by Cohen (2006), adapted to take account of the five significant domains espoused by the French film theorist Christian Metz (described in Stam (2000), p. 5 and p. 212), namely: visual scenes, visual text, speech, sound effects and music; where the concept includes the listener constructing a narrative from the information gleaned from both the visual and the auditory channels. (Ibid; pp. 892–895).

9   Filmstaden (trans. *The Film Town*) was a film studio complex situated in Råsunda, Solna Municipality in Stockholm, and the birthplace of some 400 Swedish movies. Built by the main Swedish film producer of the period, Svensk Filmindustri, it was one of the most modern film studios in Europe at the time of filming *Winter Light*. The sets for this film were unusual by virtue of having a false ceiling, which would also impact on the reverberation time of the studio (The Swedish Film Database).

10 The term Dolby 5.1 is the consumer term for the Dolby Digital cinema standard first used on *Batman Returns* (1992) (Supervising Sound Editor Richard Anderson, Re-recording Mixers Jeffrey Haboush and Steve Maslow). It contains six discrete sound channels, of which one is dedicated to Low Frequency Effects (the LFE or 'boom' channel). On cinema 35mm prints, the Dolby digital data is optically placed in line and between the film stock sprocket holes. On consumer products, Dolby 5.1 is the standard audio configuration for DVD and Blu-ray Disc players.

Although primarily associated as a 'Dolby product', the 5.1 layout was not exclusive to Dolby; commercial rivals DTS, and others, also offered products for 5.1 audio. Development of the 5.1 configuration was carried out by American Zoetrope studios, and their work actually pre-dates their association with Dolby (Murch, 2020).

5   Applying the Four Sound Areas

5.1 Introduction

This chapter describes how the Four Sound Areas framework was used during the realization of *The Craftsman* (2012) and *Here and Now* (2014) (both feature films) and Commonwealth Games Boxing (2014) (television outside broadcast).

5.1.1 Not every film director is passionate about audio post-production

Whilst the Four Sound Areas framework centres on a proposition that the emotional aspect to moving picture sound design is invaluable to the enhancement and articulation of a Director's vision, it is fair to say that not all Directors are necessarily as enthusiastic about the meticulous process by which their soundtracks are created, as they are about shooting and editing their pictures.

Sidney Lumet directed 50 studio movies and received five Academy Award nominations[1] during his career as a Hollywood film Director, an impressive number by any measure; and his 1995 book *Making Movies* is held as a reference work for those serious about film making. However, whilst Chapter 1 is entitled 'The Director: The Best Job in the World' and Chapter 5 announces 'The Camera: Your Best Friend', towards the end of the book Chapter 11 disappointingly states 'The Mix: The Only Dull Part of Moviemaking'. It opens with:

> Life has a cruel way of balancing pleasure with pain. To make up for the joy of seeing Sophia Loren every morning, God punishes the Director with the mix.

> (Lumet, 1995, p. 186)

Lumet is not the only lauded Director to hold such negative views. The French *auteur* Jean Renoir was even more vociferous in his views on post-production sound:

> I regard dubbing, that is to say, the addition of sound after the picture has been shot, as an outrage. If we were living in the twelfth century, a period of

lofty civilization, the practitioners of dubbing would be burnt in the market-place for heresy. Dubbing is equivalent to a belief in the duality of the soul.

(Renoir, 1974, p. 106)

This unequivocal aversion towards the detailed preparation work of track laying and pre-mixing the soundtrack – the paying close attention to what is reasonable to consider as half of the final on-screen product – does, however, have one distinct advantage: unlike the picture editing department, who are often working under the close supervision of the Director from early on in the post-production process, less early intervention by the Director means that the sound department is entrusted with a great deal of creative freedom as the film passes through the audio post-production stages, to be prepared for final mixing. In short, the sound department are left to get on with the job in hand, often up to the final mix; notwithstanding the creation of any interim, reel-by-reel, temporary mixes that may get viewed as work-in-progress updates.

This kind of creative freedom is espoused by Rian Johnson, the Writer and Director of *Star Wars: The Last Jedi*, who said:

Sometimes people think that a Director has a very specific vision and they're trying to get everyone on their crew to accomplish this. For me, it's always more of a collaborative journey of discovery. (American Cinematographer, February 2018, p. 45)

With that kind of journey in mind, it is hoped that the gist of the Four Sound Areas framework can form the basis for a widespread, common creative language; so that fellow professionals without an intimate understanding of the Sound Designer's or Re-recording Mixer's 'black art' – such as Picture Editors, Directors or Producers – may better understand, express and communicate their overarching intention for sound design on their production.

Sections 5.1.2–5.3.3 detail studies of commercial sound design and mixing commissions carried out on two feature films, and a live-to-air Outside Broadcast (O.B.) assignment at a major sporting event. Each utilized the Four Sound Areas as the basis for structuring their final mix. These three sound design works, by their different nature, are used to show how the Four Sound Areas framework can be adapted to any type of moving picture project.

The media for these three programmes is available for streaming on this book's companion website: www.soundformovingpictures.com.

5.1.2 *Worked example 1:* The Craftsman *(2012)*

The Four Sound Areas framework was first consciously used and appraised when I was engaged as the Production Mixer (location sound recordist), Sound Designer and Re-recording Mixer on the micro-budget short film *The Craftsman*, which was produced by Sheringham Studios as a commercial release in 2012. The screenplay is taken from a Stuart Neville short story of the same name,

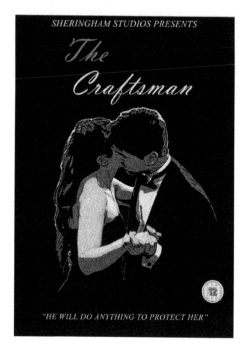

Figure 5.1 The Craftsman (2012)

which featured in the Irish crime writing collection *Down These Green Streets* (Burke, 2011).

At the pre-production stage, the Director and I began to discuss key emotional points in the script that would serve as markers for the soundtrack (four sections denoted as A, B, C and D), and I produced a simple 'sound design quadrant', to show the film's emotional high and low points; where 'positive' and 'negative' feelings would ideally be evoked in the audience, through the balancing of the sound mix.

The horizontal axis has a simple negative to positive 'scale' of emotions assigned to it and the vertical axis indicates to what extent 'reaction' or 'impact' is intended on the listening viewer, by means of the soundtrack. This was my own simplified version of the sound-mapping principle advocated by sound designer and academic David Sonnenschein (Sonnenschein 2001, pp. 18–32).

5.1.3 Plot, Sound Design and Mixing notes by section

Section A – The Hit; Albert and Celia meet for the first time

The story of *The Craftsman* explores the intense love an aging couple, Albert and Celia, feel for each other, impacted by the destructive consequences of Albert's

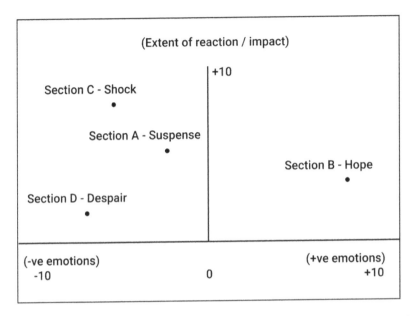

Figure 5.2 The simple Sound Design 'quadrant' used for *The Craftsman* to show the intended emotional positioning of the Sound Design for Sections A, B, C and D. The simple Sound Design quadrant of Figure 5.2 also draws on Russell's circumplex model of affect, albeit a 'semi-circumplex' version is used in this work. The vertical axis represents Valence (Russell, 1980)

haunting past. Albert is a former assassin and a devoted husband, now caring tenderly for his terminally ill wife, Celia.

Sound Design and Mixing intent: The film opens with a low-level score mixed with the sound of slow-moving tyres on gravel, as a car approaches the gates of a grand house, in a convoy with others. It pulls up at the request of the gate security guard and the driver presents his invitation. Their conversation is intentionally low in the mix, audible but unintelligible (therefore of the Abstract sound area), and the vehicle passes scrutiny. The car moves off and pulls up on the gravel drive in front of the house, and the driver steps out. We see Albert, the driver, for the first time. He takes out a guitar case and joins other alighting musicians from the convoy of cars, all carrying instrument cases.

Albert enters the building and we start to hear the hubbub of a distant yet sizeable crowd, again held low in the mix (from the Abstract sound area). The music changes tempo, quickening the pace (Temporal sound area) as Albert follows the group at a discrete distance, choosing to divert unnoticed from the party through a side door, once he is inside the building. The pace of the score again slows as we see and hear a creaky exterior door being opened, and Albert steps onto a blustery roof, the wind effects brought forward in the mix. With precise, deft movements the guitar case is clicked open and a sniper's rifle and sight mechanism is quickly assembled. Extra Foley sounds were added in post-production as punctuation and to also highlight the mechanical precision of the process, from the guitar case clips to the assembly of the weapon. Once more the music builds in tempo and level, as does the wind, the two balanced against each other and Albert's controlled breathing is now also clearly audible. All three sets of sounds – the music, the rifle assembly effects and Albert's breathing – are contributing to the Temporal area. Following Albert's ironic "Goodnight" (replaced in ADR for performance: giving greater emphasis in delivery) the music reaches its climax; and a combined silenced gunshot and glass-smash pre-empt by a fraction of a second the tinkle of a shell case hitting the floor (the Narrative sound area elements here are the line 'Goodnight', as well as the gunshot and shell case) whilst a surge in wind segues the picture mix through the Title caption and into the crowd murmur and bar piano of the cocktail party, that Albert now joins.

The gentle level and nature of this ambient, diegetic music and the general feel of bonhomie suggested by the moderate party murmur is in stark contrast to the overall level and intensity of the scene we have just left. Here the biggest contrast between the previous and present scenes is from the Temporal sound area – the urgent pace has relented and the tension carefully built up in the opening of the movie has been released by the gunshot.

We see Albert order whisky and coolly light a cigarette, and Celia enters, joining him at the bar. For intelligibility, the background crowd murmur is subliminally lowered in the mix and becomes more present on the rear speakers, as the front speakers alone deliver the dialogue between Albert and Celia. She asks him to dance and the diegetic bar piano is carefully replaced by the non-diegetic score, still played on solo piano. As a Steadicam shot slowly revolves around them dancing, the crowd murmur is faded completely away leaving just the leitmotif figure on the piano controlling the tempo of the pictures and the pair's sonically unchallenged dialogue. A photographer appears in the wider shot ready to take a portrait of the happy couple, and the background crowd atmosphere fades back in as he asks them to smile for his shot. At this point, coinciding and along with the authoritative 'pop' of the period camera's flash bulb, the sound of a gunshot (this time working in the Abstract sound area) is combined; serving as a full stop on that time period, and as a subliminal reminder to the audience of the brutal nature of Albert's work, given the intense romanticism and tenderness we have just seen between the couple.

Mixing notes: There is a noticeable Temporal change from the fast-paced opening scene (the hit) to the slow romanticism of their dance, which ends the scene.

Intended emotions: Suspense and tension: intensity is carefully built and sustained, then released by the gunshot.

(The descriptors of *Intended Emotions* here and in the rest of the chapter are taken from *A prototype analysis of emotion words.*) (Shaver et al., 1987, reproduced in Juslin and Sloboda, 2010, p. 80.)

[Footage 00:00:07 to 00:05:30]

Section B – Albert proposes to Celia

Sound design and mixing intent: This scene is designed to establish a flourishing, romantic relationship between the couple; and within the dialogue (part of the Narrative sound area) we see the cautious opening-up of Albert's emotions. He appears to be emotionally vulnerable as he strives to find the right words to express his love for Celia, and he is still guarded about the true nature of his occupation. But his hint to her of his 'cleaning' work suggests that it is only a matter of time before he trusts Celia enough to satisfy her curiosity, made obvious by her determined and repeated probing questions. Once she knows the truth, everything must change; but for now, in her state of blissful ignorance, there is still an air of innocence for her and a sense of renewal for him, following the years he has had to spend shrouded in secrecy. The prominence in the mix of birdsong (from the Abstract sound area) is an obvious presence, designed to suggest Celia's innocence to the listening-viewer and to share her feelings that all is well; whilst the presence of the sound of running water from a stream (also drawn from the Abstract sound area) is designed to allude to the figurative cleansing Albert feels in declaring his love for Celia. The scene culminates in his proposal of marriage, and her happy acceptance.

Mixing notes: the prominence in the mix of the Abstract sound area is designed to suggest a period of safety following the dangers of previous scenes; a carefree innocence and renewal.

Intended emotions: affection, happiness, peacefulness, tenderness, hopefulness.

[Footage 00:08:07 to 00:09:58]

Section C – The doctor is killed by Albert, for hearing Celia's secrets

Sound design and mixing intent: With Celia terminally ill, we see that Albert is still very much in love with his bedridden wife, being visibly kindly and gentle towards her as they talk; but old age has certainly not diminished his uncompromising resolve to retain at all costs the utmost secrecy about his past as a professional killer. The background atmosphere – mixed low against the dialogue as he dresses in their bedroom – is that of an urban summer's afternoon, drifting in through an open window: a lawnmower runs in the distance, occasionally a low-speed car drives past and the songbirds, albeit now at low-level, remain present. These are all sounds from the Abstract sound area.

Albert lightly says that he is going into town, but when pressed by Celia, he reveals that his visit is concerned with his dark distrust of the situation with the Doctor who tends to Celia; and the fact that the Doctor has heard her rambling recollections. Celia, along with the audience, is left in no doubt that Albert is going to assassinate the Doctor for what he has accidentally overheard of Albert and Celia's years as a hitman and wife. This revealing of the true purpose for

Albert's trip coincides with the absence of birdsong (from the Abstract sound area) from the mix and the introduction of a sudden, musical crescendo (from the Temporal sound area).

Later, the Abstract sound area atmosphere tracks that cover the wide shots of Albert in the park, awaiting the Doctor, are made more prominent in the mix to instil a feeling of restlessness in the audience; presented as the sound of persistent wind and rustling leaves. There are no songbirds to be heard. The coldness in Albert's heart towards the man who has been so compassionate to Celia is mirrored by the chill wind present in the front and surround channels (a utilization of the wind in the Spatial sound area). As the Doctor joins Albert, and Albert's monologue takes on a more sinister tone, the score is reintroduced in the mix with an exceptionally long, slow fade, overshadowed for much of the scene by the persistent wind in the trees. This wind is also increased in level and becomes more intense as Albert's speech progresses; the rustling of the leaves becomes noticeably sibilant, like a heavy sea wash arriving in waves on a windswept shore. The mix at this point is attempting to reflect the tumult within Albert; and through the high level of Abstract sounds, it aims to convey outwardly the turbulence and agitation he is inwardly feeling. As Albert becomes irritated by the subject of his discourse, so too the weather-borne sound effects are intentionally pushed in the mix, challenging the primacy of Albert's lines and to act as a source of irritation and unrest to the listening-viewer. This is the most crucial point of *The Craftsman*'s soundtrack. Here in the film, all Four Sound Areas are continually being balanced against each other, as their relative loudness levels exchange aural priority. The mixing of the soundtrack is designed to create dense effects, yet without completely overshadowing the intelligibility of what is being said. Towards the end of the scene, where the music is increasing in level (Temporal sound area), the exterior atmosphere (Abstract sound area) challenges the dialogue and extra Narrative sound area effects are sequentially added to the mix: a gunshot, the cries of startled birds, the flapping of their wings segueing into Albert's fleeing footsteps, and then in turn they too are swallowed by the engulfing wind and leaves, which themselves mix and resolve into the sound of running water in the sink of Celia's bathroom. The flowing water introduces the change of scene and moves the storyline back to Celia and Albert's bedroom.

Mixing notes: the complex balancing and interchanging priority of all Four Sound Areas and the building of tension by gradually increasing the overall volume of the Abstract, Temporal and Spatial areas to meet that of the constant level of the Narrative area.

Intended emotions: darkness, melancholy, anger, contempt, disgust, surprise.

The sound design intended a similar Arousal level of emotions to be evoked in Sections A and C, albeit either side of the emotional Valence axis; Albert, whether for justifiable reasons or not, is seen taking a human life. It is intended to deliver an effect that is both tense and shocking at its culmination (See Figure 5.2). **[Footage 00:10:02 to 00:14:28]**

Section D – Celia demands a promise be fulfilled by Albert

Sound design and mixing intent: In the last section of the film, the fast tempo of Albert fleeing the scene of the crime returns first to a moderate pace, and then finally to a slow tempo, as Celia herself moves slowly and painfully to bring their affairs to a conclusion. Here, just like Celia's gaunt figure, the soundtrack becomes intentionally meagre: stripped back and bare towards any of the sound areas other than the Narrative, which carries Celia and Albert's sonically unchallenged dialogue. From a mix point of view, this is the opposite of Section B's richness, where the Abstract sound area is emphasized to fill-out the soundtrack (and the surround channels). Here, Celia knows that life itself is ebbing from her, and asks that that process is completed swiftly – and in her terms compassionately – by the expert she knows Albert to be. This request from Celia of Albert, the very prospect of carrying out that act, also appears to bring a halt to Albert's own life, if only at this stage at an emotional level. But Albert's last lines, and his delivery of the dialogue, may be interpreted that he will take his own life if he is obliged to take Celia's. With the sound areas other than the Narrative dialogue sat very much in the back in the mix, the ambience is firstly an understated interior atmosphere with an occasional, almost subliminal passing car, and a score that sits deliberately low, peeping through for the most part only in the gaps of dialogue. To achieve this, the level of the music is carefully ridden in the mix throughout the scene. As the film draws to a conclusion and the couple move outside to dance one last time, the interior atmosphere is matched by an exterior counterpart, with the occasional passing cars now being the actuality of the recorded production sound. The score's leitmotif returns to the final shots of the film with a long decay of the final note allowing a punctuating 'full stop' of silence before the music rolls for the credits.

Mixing notes: Temporal change and narrower mix to focus and reflect the demise of Celia, and in turn, Albert.

Intended emotions: sadness, sentimentality, regret, sympathy, grief, sorrow, despair.

The sound design in Sections B and D are designed to evoke diametrically opposed emotions in terms of Valence, whilst of a similar level of Arousal; the first of the two scenes is based around hope, the second is grounded in despair (See Figure 5.2). **[Footage 00:14:30 to 00:26:19]**

5.1.4 An initial attempt at visualizing the mix-balance of the Four Sound Areas for **The Craftsman**

The aim of this was to roughly quantify and visualize the variations of mix-balance of the Four Sound Areas throughout the film, and to look at whether specific emotions would correspond to specific mix-balances of the Four Sound Areas.

In this initial study it was not possible to easily derive the loudness of the Four Sound Areas directly from the mix, because even though the soundtrack was designed in terms of the Four Sound Areas, *The Craftsman* was mixed at a time (2012) and on equipment before the now routine use of plug-ins, using a traditional mix-bus structure (e.g. dialogue, sound effects, music, backgrounds and atmospheres).

The use of a built-in or plug-in LUFS meter (now readily available for all types of DAWs), or even an RMS meter, could have easily provided rather more tangible level figures for comparison purposes. However, at the time that the film was created, the Fairlight Prodigy II DAW did not support plug-ins; and the subsequent lack of access to the original audio assets has made this impossible to achieve retrospectively, on a newer DAW.

Instead, the levels were subjectively quantified by me as the Sound Designer and Re-recording Mixer of the film, on the basis of the mixing notes I made at the time and the knowledge of having mixed the film. A nominal scale between 0 and 6 was used, as shown in Figures 5.3a and 5.3b, after having divided the soundtrack into relatively homogenous sections. This simple scale has '3' as being half of the perceptual loudness of '6'.

The subjective levels of the Narrative, Abstract, Temporal and Spatial sound areas were produced at five points during the long, opening scene A, one set of levels were produced for the shorter, falling in love of scene B, and at two points each for scenes C and D.

5.1.5   *Observing the Four Sound Areas at work in* **The Craftsman**

The graph in Figure 5.4 gives an overall view of the average level of sound area data on the vertical axis, for the intended emotion in each of the Four Sound Areas, across the chosen Sections A–D of the film *The Craftsman*.

The analysis of the mixing approach suggests that designing the soundtrack for negative emotions – in this case 'Suspense' (Section A), 'Shock' (Section C) and 'Despair' (Section D) – gives a similar pattern of sound area presence (a high Narrative and Temporal sound area score, but a low Abstract and Spatial one in the mix); whilst the higher the degree of audience impact/reaction I intended, the higher in perceived mixing level the sound areas contributing to these negative patterns seemed to move.

However, for the positive emotion I was attempting to evoke – 'Hope' (Section B) – a different mix pattern emerged (high Narrative, Abstract and Spatial sound area scores, but low for Temporal). Specifically, it was the Temporal and the Abstract areas that showed most change for this intended emotional effect.

5.2   *Worked example 2:* **Here and Now** *(2014)*

Here and Now is a low-budget British feature film, written and directed by Lisle Turner and made by Wrapt Films, in association with Creative England.

I undertook the sound design and full audio post-production between January and October 2013, in my studios in Moseley, Birmingham, assisted by Dan

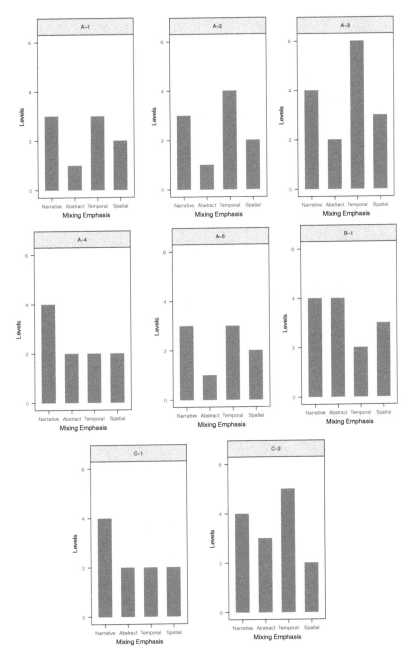

Figure 5.3a The Four Sound Areas observed through film Sections A1–5, B–1, and C1–2

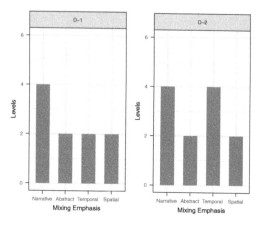

Figure 5.3b The Four Sound Areas observed through film Sections D1–2

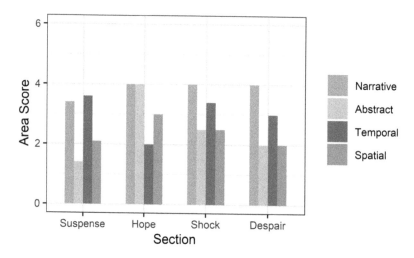

Figure 5.4 Average level by Sound Area for Sections A–D, where Section A's intended emotion = Suspense, B = Hope, C = Shock and D = Despair

Rhodes. The final theatrical mix and associated stem deliverables were created at the AIR Lyndhurst studios in Hampstead, London by me and long-standing Re-recording Mixer colleague Pip Norton.

The first public showing of the film was at the BUFF film festival on March 11th and 12th, 2014, in Malmö, a municipality in Skåne County, Sweden.

5.2.1 On sound designing and mixing Here and Now

Here and Now was shot in late summer 2012, and shortly before shooting began, I was engaged by Producer Martina Klich to oversee audio post-production, acting as the film's Sound Designer, Supervising Sound Editor and Re-recording mixer.

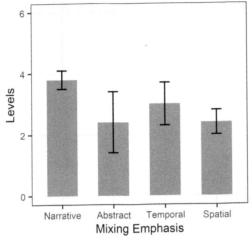

Figure 5.5 Average levels for complete film, with Standard Deviation 'σ'². The σ for Narrative = 0.3, Abstract = 1.0, Temporal = 0.7, Spatial = 0.4

In a pre-Production conversation, Director Lisle Turner discussed his aspirations for the locations he was filming in, and how he wanted distinct differences in the soundtrack for eight specific scenes.

I thought that these differences could be best achieved in audio post-production; feeling that the main consideration on location should be for the dialogue to be recorded and presented in an effortlessly intelligible state, in as natural an aural environment as possible. However (and most disappointingly) the dialogue recorded on location was of too poor a quality to be used in the final film soundtrack, requiring the replacement of all the dialogue by Automatic Dialogue Replacement (ADR), for all the film's main characters, plus two female speaking 'extras' (enabling my wife and daughter to make their feature film debuts, re-voicing these parts via ADR.)

5.2.2 *Initial sound design ideas*

From reading and discussing the script with the Director, and taking into account his desire to have eight distinct soundscapes associated with specific scenes in the film, a day spent together in my studio carrying out a spotting session, watching the first cut and listening to the Director's thoughts and comments regarding the plot and storyline, brought about a better understanding of the significance of each scene, or a particular location, shot, or line of dialogue.

What I determined during the spotting session was that *Here and Now* is not simply a boy-meets-girl love story; it is a film that was written to incorporate the

Figure 5.6 Here and Now (2014)

Buddhist principle of the 'noble eight-fold path'. Eight distinct scenes are used to illustrate each of the eight Buddhist elements, namely: *right view, right intention, right speech, right action, right livelihood, right effort, right concentration* and *right mindfulness*; and eight particular scenes within the film would each constitute a noticeable occurrence where one or both of the key characters, *Grace* and *Say*, achieved a step forward on their path towards self-discovery; and in doing so, reach an unconscious understanding of one of the Buddha's noble eight-fold path's principles.

A perceptive audience should be able to notice this progression in the two main characters over time, but to help signal this, one of my sound design suggestions was for a Buddhist bell (actually a 'singing bowl') to softly sound at the precise moment of each of the eight occurrences of insight; my *homage* towards Re-recording Mixer Clem Portman's 'angel's wings' bell used in Frank Capra's 1946 movie, *It's a Wonderful Life*. The singing bowl was used instead of a bell to carry the Buddhist analogy more obviously and to not too closely imitate Capra's original soundtrack.[3]

Other sound design considerations I suggested at the spotting session included:

• The need to establish distant farm working and a sleepily dormant agricultural feel. There is just one shot – very distant – showing a tractor working in a field late in the film that allows the use of a closer, but still distant, tractor to immediately place Grace in the countryside, e.g. her waking up in the first scene following the titles sequence. There are also several reflective, static shots of disused farm machinery. (It was anticipated that much of this would be covered by sounds from the Abstract sound area.)

- The use of contrasting bird sounds (e.g. crows versus songbirds) to help create an atmosphere of risk and danger as opposed to the rural idyll suggested by the songbirds. Differing locations being, e.g. the stark, ruined castle and the pretty flower garden of Grace's rental cottage. (Sounds such as the crows would be used predominantly in the Narrative sound area, when they would be signalling a change in mood; but may also be 'sat back' in the mix and used in the Abstract sound area.)

- Suggest the lateness of summer and the ripening of fruit in the orchard scene by adding the presence of bees and fruit flies. This scene marks a turning point in Say and Grace's relationship, as well as the turning of the seasons from summer to autumn; and the resulting sound is designed to evoke the combination of warm, bucolic days and rotting, fallen fruit. (The sounds of the insects contained within the Abstract sound area.)

- The sound of exterior wind chimes softly present on the interior atmospheres of Poppy, Say's mother's, 'hippy' barn conversion. (In the Abstract sound area, but also used for their Temporal qualities.)

- Different strengths and types of wind to induce a sense of calm, restlessness or dramatic impact. (Narrative and Abstract sound areas.)

- The sound of running water in the swimming and post-swimming scene to engender a sense of cleansing and renewal as the closeness between Say and Grace grows. (Abstract sound area.)

- Utilize the surround channels to enhance a sense of presence, immersion and perspective for the listening-viewer, particularly in the shots of Say and Grace walking through the cornfield. (Spatial sound area.)

5.2.3 Replacing the dialogue

By far the most painstaking part of the audio post-production process turned out to be the complete re-recording of the dialogue. All of the actors' lines required replacing because of the poor quality of the original location sound – primarily because of inexpertly fitted radio lavalier microphones being utilized exclusively, instead of a well-placed boom microphone in the hands of an experienced 'boom swinger'. What resulted were on-set recordings fatally coloured due to the actors' body movements creating loud rustles on their personal microphones, rendering them useless for anything other than an ADR guide track. Having to go through this process did however, in the Director's opinion, allow for an improvement of delivery, and in emotional performances, which impacted positively on the sound design of the film.

5.2.4 The application of the sound design and emotions categories

After reading the script for *Here and Now*, a development of the simple sound design quadrant (as seen in Figure 5.2) became the two-axis 'sound compass'

Here and Now

+ve reaction / impact

river

cornfield   tree house

orchard

-ve emotions _____ **+ve emotions**

cave   bridge

castle

mountain

-ve reaction / impact

Figure 5.7 Sound compass used for initial discussion with Director Lisle Turner

(shown in Figure 5.7). This was drawn to highlight where on the 'emotional spectrum' the key scenes would ideally sit. The horizontal emotions **(e)** axis runs from 'Negative emotions', like disappointment or sadness, towards 'Positive emotions', like optimism and happiness. The vertical axis shows the intended reaction/impact **(ri)** in the audience; and this represents the intended extent of the emotion attempting to be induced in the listening-viewer; in either a positive or negative context.

Whilst the Director took a keen interest in the music for the film, I was equally keen to demonstrate that it was possible in the gaps between the dialogue and music for the rest of the mix other than music to effectively convey a sense of 'being'; the Abstract sounds between the Narrative words setting the mood, place and space. From a sound design and mixing point of view, it felt important to immerse the listening-viewer and engage them through atmospheres that not only connected back to the landscape, but also brought semi-static establishing shots to life.

The chronological order, the Buddhist principle, and the positioning on the sound compass that the eight significant scenes appeared in the film is shown in Table 5.1.

5.2.5 *Plot, sound design and mixing notes*

[Times refer to the Here and Now 5.1 'Temp-mix'.]

Independent, inner-city girl Grace does not want to spend a week in the country helping her parents, Ben and Lucy, to save their marriage. Meanwhile, loner

Table 5.1 Sound design intended Buddhist principle, positioning and emotions

Scene	Narrative Principle	Positioning	Intended Emotions
1 – Castle	Right Understanding	(–e), (–ri)	Sadness, Fear
2 – Cornfield	Right Thought	(–e), (+ri)	Irritation, Contempt
3 – Cave	Right Speech	(–e), (–ri)	Sadness, Despair
4 – Orchard	Right Livelihood	(+e), (+ri)	Happiness, Joy
5 – River	Right Action	(+e), (+ri)	Enjoyment, Pleasure
6 – Bridge	Right Effort	(+e), (–ri)	Tenderness, Surprise
7 – Tree house	Right Concentration	(+e), (+ri)	Involvement, Interest
8 – Mountain	Right Mindfulness	(–e), (–ri)	Guilt, Worry, Fear

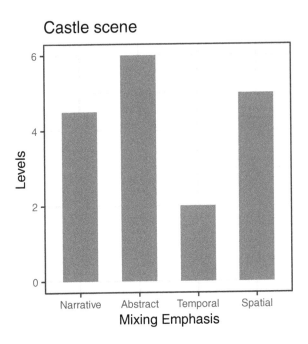

Figure 5.8a Castle scene, audio mix emphasis

and country boy Say finds Grace both obnoxious and attractive, and amongst the scenic grandeur and the dying days of summer the pair search for distraction and, against all the odds, find each other.

In an attempt to evoke emotions that would support the storyline and plot, each of the film's eight significant scenes were designed at the tracklay stage to work in a way that would go on to support the final mixing stage with an intended audio emphasis; and both the tracklay and the mixing process related directly to the Four Sound Areas of this research. The intended emphasis is shown in Figures 5.8a–h, with the relative levels noted by me whilst mixing the final soundtrack.[4]

1 – Castle: [00:13:42 to 00:16:57] (Compass –e/–ri)

Say takes Grace to a stark, ruined castle; and uncaringly she says that the village is the kind of place that if she had to permanently live there, she would jump off the castle tower. She then discovers that Say has a suicidal tendency when he climbs up the tower and appears to decide whether to jump or not. Grace, and we the audience, are shocked by this brush with death – a 'negative' emotion, inducing a 'negative' reaction to what we have just witnessed. Grace, still cold towards Say, now has the *right understanding* about him having unspoken issues.

Narrative – dialogue
Abstract – birds, wind
Temporal – score
Spatial – wind in surround channels, reverb on dialogue

2 – Cornfield: [00:20:21 to 00:25:29] (Compass position –e/+ri)

Say takes Grace into the middle of a tall cornfield, to sit in perfect peace in a bizarrely beached rowing boat. Grace, however, is quickly bored and desperate for a usable mobile telephone signal. Wandering away into the high crops, she gets

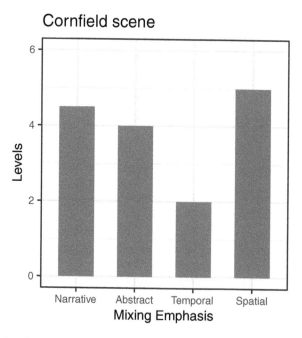

Figure 5.8b Cornfield scene, audio mix emphasis

lost and becomes frightened, calling out for Say. He finds her and she feels rescued, looking at him with the first hint of warmth and connection. Her negativity of the place and her perceived situation is countered by the positivity of her being grateful and seeing him in a different light. She has had a *right thought* about him.

Narrative – dialogue, cuckoo, mobile phone message alert, corn stalks
Abstract – birds, bugs, wind
Temporal – score
Spatial – surround panning walking through cornfield

3 – Cave: [00:31:24 to 00:36:21] (Compass –e/–ri)

Say takes Grace out of her comfort zone when he gets her to climb down into one of his special places – an underground cavern. She must place her trust in him. In the torchlight he plays *Amazing Grace* on his harmonica, and the tenderness of the gesture is not lost on her. She reciprocates that tenderness and for the first time she says a kindly word towards him. She is demonstrating *right speech*. However, the situation quickly flips, and we are left with a negative feeling about their relationship towards each other, with the disappointment that it should turn out like this again.

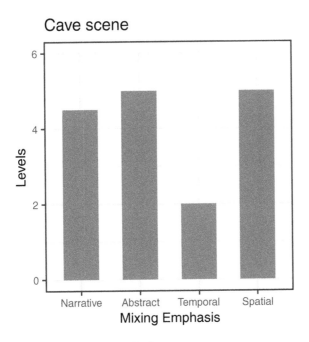

Figure 5.8c Cave scene, audio mix emphasis

Narrative – dialogue, harmonica
Abstract – room tone
Temporal – score
Spatial – room tone, emphasized reverb on dialogue and harmonica

4 – Orchard: [00:38:55 to 00:43:48] (Compass +e/+ri)

Awoken at dawn by Say, Grace joins in with the work of harvesting apples from a nearby farm's orchard. At first, she wilfully refuses to relate to, and interacts clumsily with, the other farm workers; but she comes to appreciate the serenity of a simpler country existence, living closer to nature and in harmony with the seasons. She appreciates the honest work and gains an awareness of what a *right livelihood* may consist of. There is a hopeful positivity for the viewer in this development.

Narrative – dialogue, fiddle, harvesting stick sounds, distant tractor
Abstract – birds and bugs, livestock
Temporal – score
Spatial – livestock

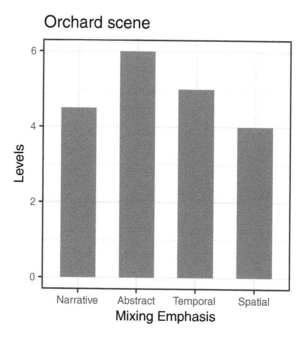

Figure 5.8d Orchard scene, audio mix emphasis

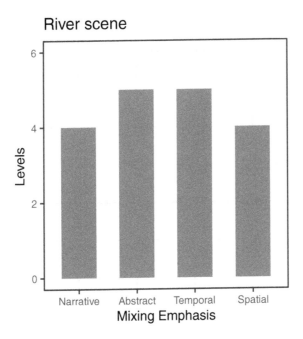

Figure 5.8e River scene, audio mix emphasis

5 – River: [00:47:51 to 00:51:25] (Compass +e/+ri)

After a disastrous day out with her parents, Ben and Lucy, Grace takes comfort in the company of Say, who surprises her by undressing on the riverbank and inviting her to swim with him. The act of swimming together, and them actually having fun, lightens her mood and brings about a feeling of intimacy towards Say. She learns how by undertaking the *right action*, she can experience new and nourishing feelings. This is the most positive and optimistic point of the film. A real change has taken place in their relationship; they are finally comfortable together.

Narrative – dialogue, swans swimming close by
Abstract – wind, trees, birds
Temporal –flowing river, water splashes, score
Spatial – wind, trees, birds

6 – Bridge: [00:52:44 to 00:54:32] (Compass +e/–ri)

Say has on-going difficulties with the village bully, Tony, and his two henchmen. Through avoiding them, and then observing their loutish behaviour from the height of the overhead bridge, Grace counsels Say to stand up to them. To show

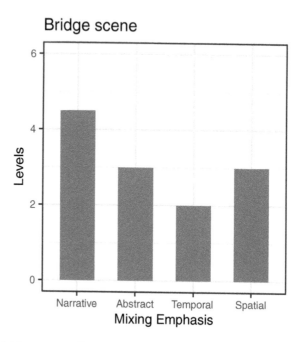

Figure 5.8f Bridge scene, audio mix emphasis

her support for Say, and her despising of the thugs, she drinks deeply from a water bottle and then spits down, unseen, onto Tony's head. As strange as this act is, it is the first time that Grace has done anything for anyone other than herself; and therefore, the intention can be seen as *right effort* on Grace's part. In emotional terms, it is positive to see Grace acting for someone else; but there is still something vaguely dysfunctional about the situation, and this strange atmosphere is left hanging in the air.

Narrative – dialogue, spitting
Abstract – leather coat creaks, wind through bridge superstructure, trees
Temporal – flowing river
Spatial – trees, wind, birds

7 – Tree house: [01:06:20 to 01:09:27] (Compass +e/+ri)

Poppy's party has been successful insofar as Ben and Lucy are now closer again, and Grace and Say are both comfortable in expressing their intimacy with each other. In Say's tree house, wrapped in a blanket and each other's arms to ward off the cool of the night, Grace wants to know about Say's future. But all too soon, Grace is called for by her father, as her parents are leaving for home. Grace has

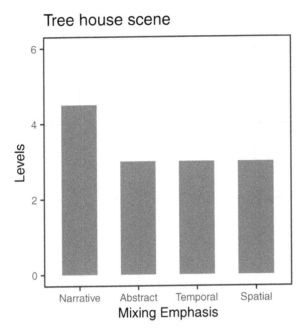

Figure 5.8g Tree house scene, audio mix emphasis

shown *right concentration* in considering, asking and listening to Say, and think-ing about someone other than herself. Witnessing this development is intended to be a significantly positive and hopeful experience for the listening-viewer.

Narrative – dialogue
Abstract – fire crackling, tree creaks, wind, owl, fox
Temporal – tree creaks, wind, score
Spatial – tree creaks, wind

8 – Mountain: [01:10:49 to 01:17:15] (Compass –e/–ri)

After his belittling and violent encounter with the bullies, Say travels purpose-fully towards the mountain, either ignoring or forgetting the arrangement he and Grace had made to meet up that morning. When he does not turn up Grace has a hunch where he will be, and she goes to the mountain after him. There, high on a ledge, we see that he is again reprising a decision that we saw in the first Act – whether to jump to his death. The weather worsens, but Grace finds him; and on the ledge, her unspoken yet loving presence brings about in him a change of heart. They have both simultaneously achieved *right mindfulness* and Say speaks freely of his father's fatal fall from the very same ledge, as though Say is released by

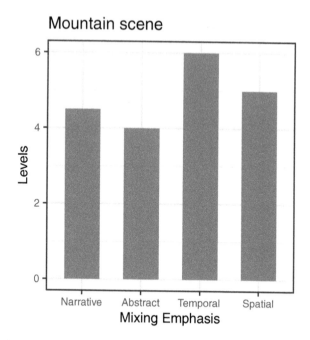

Mountain scene

Figure 5.8h Mountain scene, audio mix emphasis

her being there. In his confiding in Grace, he tells her that 'Everything changes. Everything …'[5] And we see that everything does seem to have changed for them both. Whilst not quite the most negative or positive point of the film in emotional terms, this is the closest we have seen Say to suicide; and so this scene is intended to have the most dramatic impact on the viewer, swinging from a low with Say's desperation, to a high following the scene's resolution in a kiss, that signifies that Say and Grace are ready to move on with their lives, and, metaphorically at least, escape the endless round of making the same mistakes, described by the Buddha as *Samsara* (the continual cycle of birth and death arising from mistaken concepts of self and experiences) which inevitably leads to *Dukkha* (suffering and dissatisfaction).

Narrative – dialogue
Abstract – wind
Temporal – score, wind gusts
Spatial – wind

5.2.6 Observing the Four Sound Areas at work in the mix

Summarizing the mix level values of the tables in Figures 5.8a–h, the *Narrative* and *Abstract* sound areas, over the course of the film, have equal emphasis with a mean average of 4.5, whilst the mean for the *Temporal* sound area was lowest at 3.4; the *Spatial* mean being 4.3.

Standard Deviation (σ) figures were calculated to give a measure of variability, and show that *Narrative* was the most consistent sound area with a zero deviation (it remained at a constant level) whilst *Temporal* showed the highest at 1.7, followed by *Abstract* at 1.2 and *Spatial* at 0.9.

This suggests that the *Narrative* area remained consistent throughout the film (this was predominantly the dialogue), whilst the *Temporal* and *Abstract* areas showed a tendency to 'swing' from scene to scene, usually to bring about a change in mood or pace. The *Spatial* area showed some swing, which would be consistent with the switch between sound in the surround channels and the reduction in the mix width and depth to solely the front stereo channels.

Also, within the individual scenes, the *Abstract* and *Temporal* areas are those that show the biggest change of mix emphasis within a scene; albeit at different times (*Abstract* highest in Scenes 1, the castle, and 4, the orchard; *Temporal* highest in Scene 8, the mountain).

In *Here and Now* it is the *Abstract* and *Temporal* sound areas that show the greatest variation (measured with a standard deviation (σ) of 1.2 and 1.7 respectively) from one scene to the next, suggesting that at significant plot way-points, it was the *Abstract* and *Temporal* sound areas that were mostly used in the mix to convey a change in mood or direction. The σ for Narrative and Spatial areas in *Here and Now* is 0 and 0.9, respectively.

In soundtrack terms, this often meant that the change of an atmosphere track from a close-up dialogue scene to that of a wide countryside vista was also accompanied by a change in the mood and tempo of the music.

The tendency to find the greatest emphasis 'swing' within the *Abstract* and *Temporal* sound areas in *Here and Now* is broadly consistent with the figures observed in *The Craftsman*, where the standard deviation showed that it was also the *Abstract* and *Temporal* sound areas that varied the most (σ = 1.0 and 0.7 respectively in *The Craftsman*). Which suggests (from these two films at least), that the most useful tools for a Re-recording Mixer wanting to establish a change of pace, mood or direction, lie within the *Abstract* and *Temporal* sound areas. (For completeness, the standard deviation for the other sound areas of *The Craftsman* were σ = 0.4 for the *Spatial* sound area, and 0.3 for the *Narrative* sound area.)

When comparing the patterns that emerged from creating the sound design and mix of *The Craftsman* and *Here and Now* with emotional intent in mind, it can be seen that in *The Craftsman*, the negative designed emotions – e.g. 'suspense', 'shock' and 'despair' – had the same pattern of 'sound area presence' in the mix (high *Narrative* and *Temporal*, but low *Abstract* and *Spatial*).

Yet in *Here and Now*, the equivalent suspenseful, negative emotional intent (typified by the Cornfield scene) showed its highest scores in the mix for its *Narrative* and *Spatial* sound areas. This particular *Spatial* figure may, however, be influenced by the specific nature of the scene: a head-high cornfield setting, with actors speaking as they walk through it, is as close to a perfect setting to

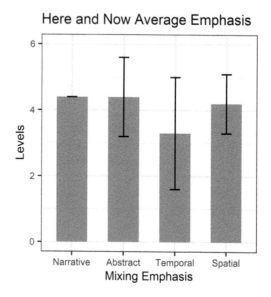

Figure 5.9 Here and Now average emphasis level and standard deviation by sound area

demonstrate surround sound as a Sound Designer and Re-recording Mixer might reasonably expect to find in a movie.

The positive emotion that was designed for *The Craftsman* ('Hope'), had a markedly different pattern to that seen for the intended negative emotions: the 'positive' emotional intent for the soundtrack showed a high *Narrative* and *Abstract* score, followed closely by *Spatial*, but with a low score for *Temporal*.

The 'negative' emotional intent for *The Craftsman* soundtrack showed high *Narrative* and *Temporal*, but low *Abstract* and *Spatial* scores; demonstrating that in this film, it was the *Narrative* and *Abstract* areas that showed most prominence when designing for a positive emotion.

However, in *Here and Now*, the positive intention designed for the river and orchard scenes showed that whilst the *Abstract* sound area remained prominent in the mix, in this film's soundtrack it was the *Temporal* area that replaced the *Narrative* area at the forefront of the mix.

5.2.7 Summary

Based on the analyses of these two films, albeit with both the sound design and mixing carried out by the same person, (me), I make two suggestions:

i) For negative emotions such as suspense, shock or despair, a Sound Designer, and in turn the Re-recording Mixer, might consider utilizing and emphasizing sounds with a highly *Narrative* nature.

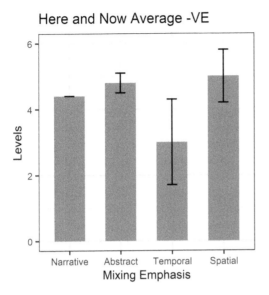

Figure 5.10 Here and Now average and standard deviations of negative emotions

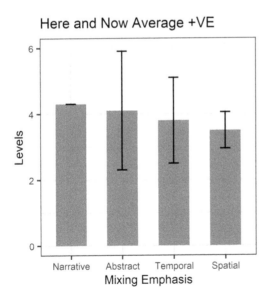

Figure 5.11 Here and Now average and standard deviation of positive emotions

ii) For positive emotions such as hope, affection, happiness, peacefulness or tenderness, the Sound Designer and Re-recording Mixer might benefit from utilizing and emphasizing sounds of a conspicuously *Abstract* nature.

5.3 Worked example 3: The 2014 Commonwealth Games

In the summer of 2014, I was engaged as the Sound Supervisor (Audio Mixer) for the two Commonwealth Games Boxing venues in Glasgow: Hall 4a at the SECC (for the appropriately termed 'knock-out' rounds) and for the finals, held at the Hydro. My responsibilities included overseeing the choice, installation and rigging of the necessary audio equipment, and then mixing the resulting broadcast audio output, synchronized to pictures, live to air. Whilst the arrangements for capturing sound at an Outside Broadcast (O.B.) event are very different to that of single-camera drama, it was of personal importance for me to determine if there were any similarities between the desired and achieved emotional outcomes for these Outside Broadcasts, compared to the film dramas of my academic research. My clear intention was to determine if a consistent approach for sound designing all types of moving picture soundtrack production was feasible.

5.3.1 Microphone logistics

The sound design philosophy and microphone layout for the Boxing finals was based broadly on that of the tried and tested rig of the heats (see Appendix 3), but with some adaptation: e.g. three stereo microphones were placed equidistant around the 270 degrees panorama that the cameras faced in the Hydro arena, arranged to cover the significantly larger audience than the heats; but the two super hyper-cardioid microphones (still designated ring 'Near' and ring 'Far' with respect to the main fixed Cameras 1 and 2 operating positions) had to be suspended from walkway gantries in the roof, which was much further away from the ring than I felt was ideal.

However, compromising their position allowed me to negotiate with the Producer and to change the commentary microphones: the commentators were finally allowed to use hand-held, lip-ribbon microphones as opposed to the 'thinner' sounding headset microphones of the heats (which were standard issue for all the Commonwealth Games sports). These headset microphones had been susceptible to picking-up large amounts of crowd and Public Address (PA) ambience 'spill', which in turn had coloured the clean commentary circuit and the overall clarity and intelligibility of the mix (as well as influencing the emotional 'feel' of the commentary itself, due to the characteristic sound of the dynamic microphone fitted to the headset's boom arm). Indeed, this change in the type of commentary microphone used was crucial in establishing the evocative nature of the Narrative sound area's exegesis speech, due in no small part to the choice of a Coles lip-ribbon microphone with its familiar, 'sound of sport' self-limiting audio characteristic.

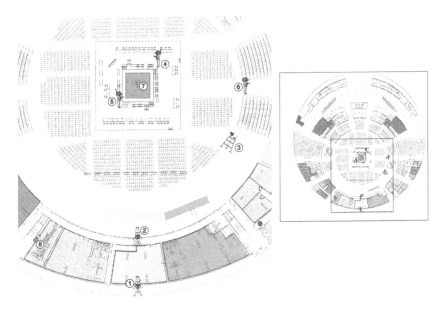

Figure 5.12 Camera plan for the Commonwealth Games Boxing finals

5.3.2 Outside Broadcast mixing and the Four Sound Areas

Starting with the heats, each of the sessions comprised 14 fights and the bouts themselves had a repetitive pattern. A rhythm soon became apparent in the nature of the mixing required; and after the first of the ten days mixing it was apparent where and when emphasis was required in the creation of an evocative, live audio mix.

Noticing which microphone faders were being 'worked' during a boxing bout, and which remained 'set', also had me thinking about the nature of the sounds being covered by each of these crowd or 'field-of-play' microphones; and by day two, the mixing desk was labelled according to what I felt were each of the console input's corresponding function in the Four Sound Areas; namely, the sources were either in the Narrative, Abstract, Temporal or Spatial sound area.

Table 5.2 shows the mixer input channel labels.

On day two, which sound areas were being 'worked' in the mix during the rounds of a bout were re-visited, to see if there was a relationship between the prominent sound area and the emotion I was attempting to convey (whilst also replicating the experience of being in the arena as a spectator, by delivering a sense of the lively atmosphere to viewers at home).

The balancing of any audio mix for moving pictures requires the Sound Supervisor/Mixer to make frequent and timely adjustments to the various audio elements; and in live television, as opposed to mixing in post-production, the

Table 5.2 Mixing console sources and corresponding sound area labels

(Sound source/microphone)	(Sound Area)
Crowd stereo wide Left/Right	Spatial
Crowd stereo Centre	Spatial
Ring Near	Abstract
Ring Far	Abstract
Red Corner	Narrative
Blue Corner	Narrative
Camera 3	Abstract
Camera 4	Abstract
Camera 5	Abstract
Commentary	Narrative
Crowd FX (group fader)	Spatial
Ring FX (group fader)	Abstract
Flash interview hand microphone	Narrative
PA High Level	Temporal
PA Vox	Narrative
EVS Red	Temporal
EVS Blue	Temporal
EVS Gold	Temporal

luxury of stopping and rewinding to allow a re-adjustment or re-take simply does not exist.

Boxing is first and foremost a contact sport and it was important to not only convey the shocking realism and power of the Abstract sound area punches (where wide-angle camera shots are used it is not possible to attribute individual punch sounds to the visuals), but also to capture the sound of those punches that are visually attributable (therefore part of the Narrative sound area), seen making contact in close-up camera angles and heard by utilizing the on-board microphones of the ring-side, mobile Cameras, 4 and 5. With these pictures, the viewers clearly saw punches landing on the boxers' heads and bodies, and heard the associated impact sound; accompanied by the excited and approving references to these sounds of combat from the Commentators (e.g. 'Just listen to the sound of those punches raining down …') at key moments in the action.

This effect of mixing the Narrative impact sounds with the more Abstract combat sounds was achieved by setting the 'Ring Near' and 'Ring Far' microphones at a specific level of amplification, held steady during the round; whilst varying the levels of the on-board shotgun microphones of Cameras 4 and 5 as they captured the Narrative sounds of gloves against gloves, heavy breathing, grunting and body blows that were clearly seen in these respective cameras close-up shots.

The required level changes between these three camera microphones, achieved through real-time fader adjustment (the physical action of 'mixing'), were large, and which of the microphones was increased in level to be the most prominent from moment to moment was dictated by which camera the boxing action was closest to. Cameras 4 and 5 were given specific sides of the ring to cover and their operators were allowed to stand just outside the ropes of the ring, at the same level as the boxers; which meant that good coverage was possible from their on-board, shotgun microphones – notwithstanding the merry dance that was being carried out by my fingers on the faders of those camera microphones.

The Abstract sound area inner-ring sounds were designed to act as a 'bed' that anchored the mix during the round; and they were used to convey to the listening-viewer at home a relentless, highly emotive sense of controlled aggression; and the violent physicality of the sport.

Whilst the Narrative impact sounds and the Spatial crowd sounds were domi-nant in evoking emotion during the rounds (and to a lesser extent, the Abstract sound area was also playing its part), the predominant sound areas in use dur-ing the breaks between the rounds and between the bouts themselves, were the Narrative (there was much more content in the breaks from the Commentators, with their analysis of the previous round), and the Temporal sound area (the PA system was playing loud, hard, fast-paced music to keep the crowd in an agi-tated state of excitement between rounds). This Temporal sound from the arena was also kept present in the television mix during the replays and slow-motion sequences of action from the previous round, heightening the agitation for view-ers at home, in a similar fashion to what the audience were experiencing at the event.

What remained reasonably consistent was the level of the faders that controlled the crowd microphones, which provided the background stereo environment for the pictures to 'sit' within and constituted the extent of the Spatial sound area. Even though the crowd got louder and quieter in the course of a bout, this was exactly the same ebb and flow, rise and fall in volume that a spectator would expe-rience sitting in the arena; therefore when making any adjustment to these levels, it was important *not* to cause either an artificially loud or quiet output level for this group of microphones, instead allowing them to 'breathe' naturally.

By labelling the faders on the mixing desk with the corresponding sound area's initial letter (N, A, T, S – see Figures 5.13a and 5.13b), it was easier to observe how the Four Sound Areas were being emphasized in the mix for emotional effect in the live transmission; and then later, from my notes, it was possible to contrast how these sound areas were utilized when mixing the live Boxing bouts, enabling a comparison to be made with the way that the Four Sound Areas were arranged in *The Craftsman* and *Here and Now*.

With the boxing matches, the 'actively mixed' sound areas of the mix during the action included the presence of Abstract sound area sounds (originating from the fixed 'Ring Near' and 'Ring Far' microphones), but with a high and consistent level of Narrative sound area content (primarily the commentary and the combat sounds from the microphones fitted to the close-up, ring-side cameras 4 and 5) and the Spatial sound area (the stereo crowd effects).

Between bouts, the use of the Abstract sound area fell dramatically in the mix and the use of the Temporal sound area rose, due to high levels of PA music being used to create excitement (and a distinct sense of disquiet) within the hall, a direct feed of which was available to me, for inclusion in the mix. This allowed the

Figure 5.13 Boxing finals mixer inputs labelled by corresponding Sound Area. Note that the Narrative sound area was initially labelled 'Logical' (L). (a) Left hand side of console, (b) right hand side of console

Figure 5.14 The Author at a Calrec Apollo mixing desk, labelled with the Four Sound Areas, mixing the 2014 Commonwealth Games Boxing finals

Temporal sound area to be carefully balanced against the Narrative area commentary, allowing the listening-viewer at home to hear the arena's atmosphere (what a spectator at the event would experience), but also hear the commentators' analysis with ease (which a spectator in the arena would not experience).

5.3.3 Summary

Although using the Four Sound Areas shorthand of 'N, A, T, S'[6] on the console certainly did not replace the need to name and group sources in the conventional way, e.g. 'crowd fx' or 'ring fx', which could be referred to as '*Primary labelling*', the use of the Four Sound Areas nomenclature was extremely helpful as '*Secondary labelling*'.

Separating the two halves of the challenge for an Outside Broadcast Sound Supervisor/Mixer in this way seemed to give more clarity and focus; and provided a visual feedback to better understand and check what was being arranged 'emotionally' within the live mix.

Using the Four Sound Areas as a basis for the sound design and the mix of a live-to-air sports assignment translated readily and easily to the task; and operationally, the framework functioned well. Labelling the mixing desk inputs with their corresponding and appropriate sound area was helpful in one particular aspect – it allowed the sound sources to be grouped simultaneously, in both a technical and a creative context; allowing an easier separation of the technical *necessities* (e.g. such as 'are the microphones working?', 'Can a signal get from one place to another?') from the *niceties* of delivering an emotional experience to the listening-viewer at home.

When I compared the sound design quadrant used for the intended emotions in *The Craftsman* (Figure 5.2) and its successor, the sound design compass used for the intended emotions in *Here and Now* (Figure 5.7), what was intended to be emotionally evoked by the boxing coverage bore closest similarity to three specific scenes in *The Craftsman* (scenes A, C and D – 'Suspense', 'Shock' and 'Despair' respectively) and one scene in *Here and Now* (the Cornfield scene).

And as shown earlier, the two suggestions that presented themselves from the analysis of *The Craftsman* and *Here and Now* were:

i)   For negative emotions such as suspense, shock or despair, utilizing sounds with a highly *Narrative* nature in the mix worked well.
ii) For positive emotions such as hope, affection, happiness, peacefulness or tenderness, the mix benefited from utilizing sounds of a highly *Abstract* nature.

As the Sound Designer/Mixer, it would seem that my intended emotional effect for the sound design of the Boxing was for it to be, overall, emotionally 'negative'. If the earlier findings of *The Craftsman* and *Here and Now* were consistent, this should therefore mean that the sound area's most prevalent in the mix should either be the Narrative and Temporal sound areas (as in the case of *The Craftsman*) or the Narrative and Spatial areas (as in the case of *Here and Now*).

In fact, both situations were observed as each boxing bout moved between two distinct states: during the rounds (the 'action') and during the inter-round interval (the 'respite').

During the action phase of the round itself, the predominant sound areas at work in the mix were sounds from the Narrative and Spatial sound areas – e.g. Narrative: commentary, referee, timing bell, timing blocks and close-up punches; and Spatial: the crowd.

(Referring back to the feature film soundtracks, the Narrative and Spatial areas were used to create *Here and Now*'s negative characteristics for the Cornfield scene.)

The Abstract sounds of movement and general combat, e.g. the boxers feet on the canvas ring flooring, non-specific punches and grunts were also present, but they were much less prominent in the mix than those of the Narrative and Spatial sound area; whilst the Temporal sound area sounds were seemingly not engaged.

However, in the respite phase between the rounds, the Abstract and Spatial sound areas were greatly reduced in the mix; and it was the Narrative and Temporal sound areas that were the most prominent – e.g. Narrative: the Commentator's analysis, overheard instructions to the boxers from their trainers and seconds[7] in the corners, the signal of the start-of-round bell; and Temporal – the high-energy music played loudly to the crowd from the arena PA; the same sound areas that were used to create negative emphasis in *The Craftsman*.

5.4.1 Discussion points

As described in this chapter, the entire dialogue for *Here and Now* was replaced in the studio after filming was complete, using the ADR technique. American films often replace a film's entire dialogue with the same actors' voices; whilst Italian films routinely replace the on-screen actors' voices in ADR with differ-ent actors' voices[8] (to differing contentment of the domestic audience).[9] French cinema, like the British film industry, has a proud tradition of recording 'usable' location dialogue.

- Is there an artistic untruth being presented to an audience by the use of ADR, compared to well-recorded location sound capturing the original performance?
- Contrast and describe the style and tone of dialogue presented in the pre-dominantly location recorded voices of *Côte d'Azur* (2005) with the close-presence style of dialogue that uses ADR in *Garden State* (2004).

5.4.2 Practical exercises

- Record a passage of dialogue out of doors, simultaneously on a personal lavalier-style microphone and a shotgun boom microphone, recording syn-chronized pictures of the actor, or actors, speaking. Compare the differences in tone between the two types of microphones.
- Record the same passage of dialogue with the same microphone set-up indoors. During this recording, have the actors attempt to match the mouth movements of the original pictures. Compare the differences between the microphones again.
- Which recording (outdoor lavalier or outdoor boom; or indoor lavalier or indoor boom) best serves the artistic intention of the original outdoor pictures?
- Repeat for a suitable indoor filming scenario.

Notes

1 Sidney Lumet (b. 25/06/1924 – d. 09/04/2011) is described in *The Encyclopaedia of Hollywood* as one of the most prolific filmmakers of the modern era, directing on aver-age more than one movie a year from his directorial debut in 1957 (Siegel, 2004, p. 256). Interestingly, although he received five Academy Award ('Oscar') nominations over a slate of 50 films, he never won an Oscar for his film craft or artistic work. He did, however, have an Academy Honorary Award bestowed on him in 2004 'in recogni-tion of his brilliant services to screenwriters, performers and the art of the motion pic-ture'. His soundtracks, however, judging from his textbook on filmmaking, remained consistently low on his list of priorities throughout his career.

2 Sample Standard Deviation (σ) is being used here as a measure of the conspicuous loudness variability throughout the film. The value of the standard deviation is shown by a line superimposed on each of the four sound area columns. It may help to visualize this line as a representation of how far a group fader for each sound area would have moved up and down throughout the film.

3 Director Frank Capra insisted on recording all of the actors' dialogue live on location for *It's a Wonderful Life*, which presented a particular audio challenge: artificial snow was required to cover the set as they filmed, mid-summer, on a San Fernando Valley studio backlot. Traditionally, artificial snow for filming was made from breakfast cereal cornflakes painted white; but the actors' footsteps on the cereal would have interfered with the sound recording, and so a new type of 'faux snow' was created from water, soap and fire-fighting foam (Hickman, 2011).

4 I still routinely use a Peak Programme Meter (PPM) as a reference tool when mixing and this was used as a convenient and simple scaled guide when noting the average levels of the four sound areas, across the individual scenes, as mixing took place.

5 The revered Zen Buddhist teacher Suzuki Roshi (b.18-05-1904 – d. 04-12-1971), when asked to summarize Buddhism in a simple phrase, said: 'When you realize the truth that everything changes and find your composure in it, you find yourself in Nirvana' (Kornfield, 2014, p. 57).

6 Initially, the Narrative sound area was referred to as the Logical sound area. It can be seen denoted 'L' in the labelling of the sound desk in Figures 5.13a and 5.13b.

7 The cry of 'seconds away' often heard before the start of a bout, or the start of each round, does not refer to the time remaining before the fighting begins; a Second is the boxer's assistant, providing water, applying grease to facial cuts, a fanned towel for cooling and a stool for them to sit on in the ring.

8 Dictator Benito Mussolini banned the use of foreign languages in films shown to Italian audiences in 1930. The law was rescinded after the Second World War, but by then it had become a part of the popular culture (Terranova, 2014).

9 The Italian film industry tradition of 'dubbing' English language films into Italian became so unpopular that DVD and Blu-Ray releases in Italy were forced to include the original language, as well as the Italian version (Zanotti, 2015).

6   Reflections on practice

Introduction

In this chapter I offer self-critique on the three pieces of work discussed in Chapter 5, compare the differences in the use of the Four Sound Areas in those practice pieces, present comments from film critics relating to the movie *Here and Now* and discuss the differences and considerations when creating a sports Outside Broadcast mix, compared to a drama mix.

6.1 The sound design and mix for *The Craftsman*

Central to the mixing philosophy of *The Craftsman* was the desire to provide 'effortless intelligibility' for the listening-viewer with regard to speech. However, as stated in Chapter 5, it is important to emphasize that dialogue does not necessarily constitute the entire Narrative sound area of a film; in the same way that the Abstract sound area does not only equate to sound effects.

In *The Craftsman*, one specific part of the Narrative sound area – the actors' dialogue – remained monophonic throughout by intent; to provide a greater degree of separation in the mix from the atmospheres used within the Abstract sound area and the music used within the Temporal sound area, affording it a clear avenue of direct communication.

The *amount* of sound from the Narrative, Abstract, Temporal and Spatial sound areas in use at any one time changes frequently in a mix, as these Four Sound Areas are continually re-balanced against each other by the Re-recording Mixer; and the decisions behind which sound area has precedence over another at any particular moment of the mix, means that the subjective *balancing* of the four soundtrack elements has a significant impact on what might be considered as the overall sound design.

A challenge exists in how to graphically represent these instantaneous and moment-to-moment changes, as a soundtrack differs from a picture track by virtue of it being silent at any moment of a freeze-frame; the sound itself can only be auditioned by it moving – by the soundtrack actually playing. One solution was for the full soundtrack to be broken down into its four separate sound areas,

Figure 6.1a The Craftsman Section A: Narrative, Abstract and Temporal areas

whose audio stems could then be output as four individual waveforms, which in turn, could be examined and compared over a common timeline.

Figures 6.1a and 6.1b demonstrate such an arrangement for Sections A and B of *The Craftsman* soundtrack. In each figure, the top stereo track is the waveform of the Narrative sound area (I had initially called this the Logical sound area), the middle stereo track is the Abstract sound area and the bottom track is the Temporal sound area. (Note that the Spatial sound area has been intentionally omitted for clarity in Figures 6.1a and 6.1b, as the waveforms show little variation for the two chosen scenes.)

The objective is to give an impression of the relative *balance* of the sound areas, and these screenshots show interesting variations in the Narrative, Abstract and Temporal areas (the top, middle and bottom traces, respectively) between the dynamic opening of Section A and the serenity of Section B.

Another viewpoint is provided by the Fairlight Prodigy II audio editor screen-grabs (Figures 6.2a and 6.2b), which show the contrast between Sections A and B; this time the difference is in audio clip density: it can be clearly seen that more numerous, individual sound clips were used to create the soundtrack in Section A than in creating Section B. Table 6.1 shows this track layout.

Figure 6.1b The Craftsman Section B: Narrative, Abstract and Temporal areas

These screenshots of the DAW's audio editor display (Figures 6.2a and 6.2b, with the active audio tracks listed in Table 6.1, may not necessarily have direct relevance, but it is interesting to show and compare the clusters of individual clips that make up the soundtrack, at two different emotional points in the film.

What is interesting to note with Figures 6.2a and b is the contrast in the density of the clips, e.g. the number of sound clips used to evoke emotions in a scene calling for tension and suspense (the overriding emotions designed for Section A) compared to the number of sound clips used to evoke emotions in a scene portraying peacefulness and serenity (the overall intended emotion for Section B).

Perhaps the difference in the number of sound clips used in Section A, compared to Section B, suggests that a Sound Designer must work harder in terms of the number of sounds required to support an evoking of tension and suspense, than in evoking peacefulness and hopefulness. The extra concentrations of clips in this example occur predominantly in the Narrative and Abstract areas of Section A, leading up to 'the kill' before the release of tension occurs.

This release of tension in the scene is also identifiable in Figure 6.2a, as a peak or 'spike' in level for both the Narrative and Abstract sound areas, and as an evolving crescendo in level for the Temporal sound area, occurring at the end of the first half of the opening scene.

Figure 6.2a Dense soundtrack for Section A [duration 05' 24'']

One other interesting result from plotting the relative levels of the sound areas at various points throughout the film is that the Abstract sound area appears to only make a significant impact on the soundtrack in Section B. Given that the film throughout addresses deep emotional issues, this may be surprising. What was it that set Section B apart from the other sections of the film?

During Section B, the soundtrack is markedly different in what it attempts to convey emotionally compared to the other sections: e.g. affection, happiness, peacefulness, tenderness, hopefulness; and indeed, Section B does sit as an outlier on the Sound Design compass. It is alone in being positive in both the intended reaction/impact on the listening-viewer and its planned steering of their emotions; the sound design in accord with the Director's ideal of this being the only self-contained scene of hopefulness and renewal – one that provides respite from the otherwise omnipresent shadow of death, inferred by Albert's trade and by Celia's terminal illness.

6.2　Critical appraisal and sound design notes for *Here and Now*

In the UK, *Here and Now*'s release was met with critical acclaim: it was named by cinema writer and critic Mark Kermode as his 'film of the week' in the July 4, 2014, edition of the BBC Radio 5's *Kermode and Mayo's Film Review* show; and Kermode's review directly addresses the soundtrack:

Figure 6.2b Much sparser for Section B [duration 02' 06''] (See also Table 6.1)

The film has wonderful pastoral imagery and there's also a nice musical motif that goes on. At the beginning she [Grace] has these headphones in, it's all urban noise; and as she gets out into the countryside you start to hear people playing fiddles in bucolic under-tree settings and there's a moment when he [Say] takes her to a cave and he plays a harmonica so she can hear the sound of the acoustics echoing. (Kermode and Mayo, 2014)

And he reprises his approval for the film in the July 6th, 2014 edition of *The Observer* newspaper, again referring to the mixing of the soundtrack:

As the headphones come off and real-life filters in [...]

(Kermode, 2014)

Whilst Peter Bradshaw writing in the *Guardian* on July 3rd, 2014 described it as:

An intelligent and attractive film. An impressive debut feature from Writer-Director Lisle Turner that's well acted. (Bradshaw, 2014)

Table 6.1 Fairlight Prodigy II track layout for mixing *The Craftsman*

Channel	Stereo/Mono	Section A	Section B
Music 1	Stereo	ON	
Music 2	Stereo		ON
Atmos 1	Stereo		ON
Atmos 2	Stereo	ON	ON
Atmos 3	Stereo	ON	ON
Atmos 4	Stereo		
Dial 1	Mono	ON	ON
Dial 2	Mono	ON	ON
Dial 3	Mono	ON	
Dial 4	Mono		
Reverb Dial	Stereo	ON	
Reverb FX	Stereo	ON	
Stereo FX 1	Stereo	ON	
Stereo FX 2	Stereo	ON	
Stereo FX 3	Stereo		
Stereo FX 4	Stereo		
Stereo FX 5	Stereo		
Stereo FX 6	Stereo		
Mono FX 1	Mono	ON	
Mono FX 2	Mono	ON	
Mono FX 3	Mono	ON	
Mono FX 4	Mono	ON	
Foley 1	Mono	ON	
Foley 2	Mono	ON	
Foley 3	Mono	ON	
Foley 4	Mono	ON	
Stereo Layback	Stereo		

Total Film magazine's reviewer James Mottram also approved, writing:

> Lisle Turner manages to capture both the rhythms of rural life and the sharp
> pangs of adolescence with a quiet assurance. Modest but moving. (Mottram,
> 2014, p.53)

The purposely rustic and simplistic sound design, carried through to the mix,
would therefore seem to have worked overall, and it serves the pictures well;
but it occurs to me that it may be said that the soundtrack functions primarily
by virtue of only two of the four available sound areas, throughout most of the
film.

For instance, the dialogue remains at a constant perceived level and stays
firmly in the Narrative sound area throughout the film; and the music score *almost*
never leaves the Temporal sound area (two exceptions being when it is used in
the Abstract sound area, as a mysterious beginning to the Cave scene as Say and

Grace approach and then climb down into the cave; and then immediately afterwards, inside the cave, the music takes a place in the Spatial sound area, as Say's rendition of *Amazing Grace*, played on his harmonica, reverberates around the cave).

Whilst these two sound areas alone (the Narrative and Temporal) can – and do – tell the bare story, they do so in almost complete isolation from the images on-screen; presenting what might be described as a form of sonic *tableau vivant*. The Narrative and Temporal sound areas are detached from the landscape that both underpins the story's visual environment and highlights the difference between the worlds that Say and Grace, metaphorically, inhabit.

These two sound areas alone provide what is necessary to fulfil a narrow, although indisputably important, role of the top-level descriptive function; but put simply, in this film it could be said that the Narrative and the Temporal sound areas are being used in the mix to tell the listening-viewer *how* to feel, by limiting the range of emotional intent being presented in the Narrative sound area, along with reduced emphasis on Abstract sound area sounds.

Yet the contextual and emotional engagement with the plot – a deeper level of *what* the Sound Designer and Re-recording Mixer hopes a listening-viewer will feel – is left in the hands of the *Abstract* and *Spatial* sound areas, e.g. the archetypal surround-sound example of walking through a cornfield, with head-high stalks swishing and parting around and behind, immersing the listening-viewer with content in the surround channels. Or the tiny activity sounds of the wasps, bees and fruit flies that are present in the orchard scene to place the picking scene – and therefore the listening-viewer – at a particular, and distinctive, time: the end of the summer fruit-growing season in England.

On the one hand, the Narrative and Temporal area sounds in the mix *prescribe* the emotional experience; on the other, the Abstract and Spatial area sounds encourage a more *personal interpretation* of what is being presented to the listening-viewer on-screen.

6.3 Labelling the console

Arranging and labelling the console in line with the Four Sound Areas when mixing either *Here and Now* or *The Craftsman*, would make for a far less straightforward layout, compared to the Four Sound Areas layout that was used for mixing the Commonwealth Games O.B.

As opposed to delivering the narrow sonic soundscape of a sporting occasion, such as a football, cricket or boxing match, a filmed drama encompasses a much wider range of situations and places that fictional characters inhabit (or more accurately, that through the sounds utilized by the Sound Designer, the listening-viewer will believe them to inhabit). This results in many more individual sound cues being used in a drama mix to evoke, suggest, and reinforce an emotional engagement with on-screen events and circumstances, compared to the number of sound sources utilized when presenting sports coverage.

6.4 The treatment of the Outside Broadcast sport mix compared to mixing the film dramas

In the cases of *The Craftsman* or *Here and Now*, it would not have been logistically advantageous to secondary label the mixing desk's channel faders in the same way as the Outside Broadcast desk, with the designators N, A, T and S (Narrative, Abstract, Temporal and Spatial) used to signify the sound area that they routed. This, I would suggest, is because in a sporting context each sound *source* is usually fulfilling one distinct function within the framework of the Four Sound Areas. Whereas in a drama situation, sounds are much more likely to move freely between sound areas in the Sound Designer's quest to evoke audience emotion; and they arrive at the console from many more, multiple-track, sources.

One other explanation for the difference in the usefulness of secondary labelling an O.B. console with the Four Sound Areas, but not so a drama mixing desk, could lie in the concepts of *Narrative Fit* and *Functional Fit*, which are often seen working together in Gaming sound (Ekman, 2014).

Applied here as a comparison between the fictional dramas of *The Craftsman* and *Here and Now* and the Commonwealth Games boxing, the dramas may be considered to utilize *Narrative Fit*, whilst the sporting event requires a soundtrack that satisfies *Functional Fit*.

In drama, the sounds used to deliver an enhanced understanding of the story – the Narrative Fit – can often produce a strong source of emotion in the listening-viewer. Ekman suggests:

> The most pivotal sounds in film are typically refined to give the sound extra emphasis, loading them with narrative, connotative and symbolic meaning, and enhancing their attention-grabbing effect. (Ekman, Ibid.)

Conversely, comparing sporting event sound with that of the Functional Fit sound utilized in gaming, may not at first seem like a likely match; but the bringing together of simple, discrete sounds, that are then grouped together to form the whole soundtrack – as in the direct use of Narrative, Abstract, Temporal and Spatial sound area sources – and labelled as such on the mixing console (as in the case of the Outside Broadcast sport example), does suggest a relationship to Ekman's thoughts on Functional Fit; e.g. sounds that will be frequently repeated, such as a referee's whistle, a starting gun, a timing buzzer or bell, or an Umpire's voice delivering the match score. Ekman expands this point:

> The driving factor for creating functional fit is to consider the utility of sounds for play [...] sounds are overly simplified, grouped together, and tend to match game actions categorically with earcons[1] or auditory icons.[2] (Ekman, Ibid.)

Overall, the sound design for the Commonwealth Games boxing provided the kind of aural experience that the Production team demanded, and the mixes delivered the emotional experience that I was looking to convey.

However, given more access and time to place extra microphones in and around the field of play would have undoubtedly provided me with the opportunity to further enhance the sounds of combat from within the ring; with the possibility of an associated increase in the extent of the emotions experienced by viewers, such as a larger involvement and interest on the positive side, or greater disgust and repulsion on the negative side, through more clearly hearing the sounds directly associated with the shockingly violent nature of this sport (and depending on the viewer's own stance on boxing).

This approach to coverage, however, would almost certainly have resulted in an experience too graphic for a general television audience, and therefore unsuitable for the kind of broadcast coverage required for the Commonwealth Games.

But this kind of enhanced coverage is at least real; and distinctly different to the practice of enhancing the sounds of a sport artificially.

In the televised rowing events at Sydney 2000, the Olympic Broadcasting Services Head of Sound Dennis Baxter played-in the sampled sounds of oars cutting through water, along with boat movement and crew murmur recorded from earlier, un-televised practice sessions, which were free from the noise and colouration of engines from motorboats and helicopters that were being utilized as mobile camera platforms (Sullivan, 2012). Baxter defends this policy:

'Some people think it's cheating. I don't think I'm cheating anybody,' he says. 'The sound is there. It is the exact sound. It's just not necessarily real time. Because of the laws of physics, you've got one noise masking another noise. So ... When you see a rower, your mind thinks you should hear the rower and that's what we deliver.' (Ibid.)

Baxter's approach to sound design in this example – artificially representing the sounds of oars on live sport footage – can be seen as an example of him making an emotion-based sound design decision: he is delivering what the listening-viewer *expects* to hear, as opposed to what they would *actually* hear, in real-time from microphones covering the event. And whilst Baxter's approach may be no different to that of mimed musical performances on pre-recorded television shows, what can be argued is that this clearly demonstrates an approach to sport sound design that is expressly intended to enhance the listening-viewer's emotional experience. Through the clearest aural representation of what is being shown on-screen, the Sound Designer is creating an emotional experience that includes feelings of greater involvement and interest, as well as heightened enjoyment and pleasure.

This is precisely what UK broadcasters such as BT Sport and Sky Sports are keen to promote to their subscribers – at a premium price – through the introduction of Dolby Atmos soundtracks, alongside their Ultra High Definition (UHD) '4k' pictures (Dziadul, 2016; Langridge, 2017; Martin, 2017).

A previously unconsidered use of pre-recorded crowd effects occurred when the relatively small football club West Ham United moved to the London 2012 Olympic stadium in 2016. Given the small number of fans in such a large ground, as the club struggled along in the second-tier Championship league, an attempt was made to artificially boost the atmosphere within their new home by playing the sound of a large and excited crowd over the PA system. It was not received well by the fans (Dream Team FC, 2016; Freebets, 2016).

But given the requirements for football games to be televised from behind the 'closed doors' of silent stadia, empty of fans either for disciplinary reasons (Associated Press, 2019) or as a result of the worldwide social-distancing measures imposed by governments, following the outbreak of the 2020 COVID-19 pandemic that impacted on spectators attending sporting events globally,[3] it has surely been proven beyond any reasonable doubt that pictures alone fail to create the heightened sense of excitement that television audiences have come to expect.

And sports professionals would also seem to agree on the fans' presence being an integral part of the game – as Manchester City manager Pep Guardiola put it when addressing the importance of the crowd's involvement:

> you have to ask is it worse to play football without the spectators[?] We do our job for the people and if the people cannot come to watch us, there is no sense. (Gonzalez, 2020)

Meanwhile, England cricketer Jofra Archer suggested that crowd sounds *should* be played-in at cricket matches, when the players faced the prospect of performing to empty grounds as a result of the government-imposed COVID-19 safety measures:

> We play music at cricket. Why can't we play some crowd simulation? We can play the clapping, play the oohs and aahs and just try to make it as realistic as possible. (Wigmore, 2020)

However, companies televising cricket resisted the opportunity to add pre-recorded crowd effects when Test Matches, One Day Internationals and T20 Cricket resumed, spectator-less, in the summer of 2020. But many other sports took the opposite view, and decided in favour of utilizing artificial crowd noise, to mask the obvious absence of spectators: broadcasters of diverse sports such as Australian rugby, Korean baseball and the American NFL all played-in the sound of crowds whilst televising events, with live football on TV from the English Premier League, Germany's Bundesliga and Spain's La Liga drawing on the EA Sports *FIFA* series of gaming soundtracks to provide authentic match atmosphere beds and effects, from the extensive *FIFA* sound library (Keh, 2020; Shergold, 2020).

But as an example of exciting sports sound design, it is fair to say that for European football at least, even with the inclusion of these *FIFA* crowd sounds, the mixes were disappointingly poor across the board: the lag between an on-screen event and a corresponding audio cue was consistently too far behind play for instance; and the general crowd atmosphere track was little more than a vague 'wash', set unnaturally low in the mix. The unfortunate (and obviously unintended) impression was that of a half-hearted attempt by sports broadcasters to satisfy the viewer's need for a crowd's 'presence' – yet the resulting soundtracks consistently missed the opportunity to fully embrace these artificial sounds as an integral part of the programme's audio attributes.

Noticeably, the difference in the emotional 'feel' between such dull, live match sound, compared to the exciting, post-produced 'sizzler' highlights for web consumption (which did take advantage of well-balanced, pre-recorded crowd reactions) was strikingly obvious. In a post-Coronavirus world, it will be interesting to see if or how the use of pre-recorded sound effects in live television sport evolves.

Baxter broke the mould in the Sydney Olympics by faking the actuality coverage of the event taking place; but the blurring of such a boundary, in the name of enhancing the listening-viewer's emotional experience, comes down to deciding whether the objective of sound for sport is to deliver reality, enhanced reality, or to embrace the stylized concept of Baxter's falsified sounds. Each has its own merit, morals (or even necessity); but remain testament to the implicit need of sport to include conspicuous, emotion-based sound design.

Two other observations can be made on the differences in sound designing for emotional effect at a sporting event, compared to a feature film.

Firstly, the sports Sound Designer/Mixer is usually looking to only convey a single overarching emotional experience to the listening-viewer watching the event: a sense of heightened reality combined with an authentic representation of actually being at the event.

In doing so, this not only involves reporting the activity taking place on screen but also the experience of being amongst other people in a stadium setting; which, depending on the sport, can also include a privileged access to the kinds of sounds usually only heard by the participants on the field of play: such as the players' cries to each other, the striking of the ball or the referee's observations. This narrow objective is in contrast to a gamut of possible emotions that could be presented for personal interpretation by the audience of filmed drama, throughout its soundtrack.

Secondly, that sound of the audience at a sporting event – the crowd's reactions to the state of play – has shown itself to be an integral part of delivering the emotional experience to television viewers. Conversely, the sound of the audience's reaction to a cinematic presentation does not function as a part of the sound design (even though it can add to the communal experience of watching in a movie theatre).

The factors that influenced the mixing of these three practice pieces might be considered as so different in nature that the objectives for the *Craftsman* and *Here and Now* soundtracks are somewhat at odds with that of the Commonwealth Games boxing Outside Broadcast.

But whilst it is the case that creating and mixing soundtracks for a live sporting event is fundamentally different to the process of mixing sound for fictional drama, as far as creating and mixing sound is concerned, what is central to the principle of the Four Sound Areas is the ability to achieve enhanced emotional engagement with an audience; be it the 'realities' of covering a sporting event, a documentary hoping to bring about social change, or even within 'Reality TV' (where the filming of ordinary people going about their ordinary business is viewed as being more akin to entertainment, than it is about delivering of information about the subjects). Filmed drama meanwhile works to create a sense of reality from a fictional subject.

Therefore, the congruency between the soundtracks of all three pieces comes from the use of the Four Sound Areas framework, and from the desire to deliver enhanced emotional experiences for the listening-viewer.

6.5.1 Discussion points

- Should Outside Broadcasts cover the 'actuality' of field of play sound as faithfully as possible, or is it an opportunity for the use of cinematic, exaggerated effects?
- Which sports would be enhanced by the use of overemphasized or enhanced sound effects? Which would be diminished by it?

6.5.2 Practical exercises

- Use multiple microphones to record the sound of a tennis game. Consider the stereo image. Compare how recording on an indoor court differs to recording on an outside court. Compare the sound of the microphones in isolation with the sound of them balanced together.

Notes

1 An earcon is a sound of short duration, usually musical in nature such as a 'beep', which indicates or communicates that an event has taken place, typically within a computer system. It is a play on words from the visual term 'icon', and its pronunciation in English, 'eye-con'.
2 The term auditory icon was coined by Bill Gaver in the early 1980s during his research into the use of sound for Apple's file management application 'Finder'. Similar to the earcon by providing feedback in computer-based user interfaces, the auditory icon uses everyday sounds to denote events and actions. The term has become commonplace in the shifting world of digital and is deemed the sonic equivalent to visual icons seen within most operating systems (Gould 2016).
3 In 2020, the COVID-19 pandemic resulted in an unprecedented impact on sporting events around the world, including the postponement of the Tokyo 2020 Olympics, the Paralympics and the Euro 2020 tournament; the cancellation of the entire NBA basketball season, the suspension of major league football (soccer) throughout the world, and the postponement of Formula 1 racing, the Wimbledon Tennis tournament and the Open Golf championship.

Part 3

Communicating the soundtrack

I love creating partnerships; I love not having to bear the entire burden of the creative storytelling.

Steven Spielberg

7 Adding to the lexicon

Introduction

In this chapter I look at how the original question I set out to answer was addressed.

7.1 Answering the question

That original question that I asked was:

> Can the emotional impact of a film be affected by balancing the mix of the sound design elements in specific ways?

Which resulted in me asking a further question:

> Can a practical framework be developed to help facilitate the induction or enhancement of specific emotions in an audience via the mix-balancing of sound design elements by the Re-recording Mixer?

As a result of attempting to answer these two questions, I proposed a practical sound design structure, and called it the Four Sound Areas framework. In this framework, any moving picture soundtrack is considered in terms of its Narrative, Abstract, Temporal and Spatial components.

I further proposed that the mix-balancing of these Four Sound Areas can influence the emotional impact of a film or audio-visual work for an audience.

This framework was used, both conceptually and practically, to mix the soundtracks of productions analysed in this book: a short film *The Craftsman*, a full-length feature film *Here and Now* and to steer the mixing of live, televised Boxing matches at an Outside Broadcast event.

A subjective analysis of the practice was carried out in order to reflect on the impact of using this framework to facilitate emotional sound design; and the main results I found from using the Four Sound Areas framework on these practical examples were:

i) The Four Sound Areas framework is simple and straightforward to integrate into the established and existing workflows for audio post-production.

ii) It is equally suitable for use by Sound Designers (e.g. sound editors whilst preparing soundtracks for later mixing – i.e. a 'bottom-up' *tracklay*) and sound mixers (e.g. a Re-recording Mixer receiving the output of a Sound Designer/Supervising Sound Editor – i.e. a 'top-down' *mix*).

iii) Mixing emphasis is more easily arranged by connected sounds being conceptually grouped into the Four Sound Areas of Narrative, Abstract, Temporal and Spatial.

iv) The Four Sound Areas can act as a useful framework for Picture Editors to more easily consider emotive sound design at the picture editing stage; enabling the audio elements of a timeline to be arranged and structured in a more cohesive way for the audio post-production stage.

v) It is also a useful tool for informing the academic examination of soundtracks for both analytical and aesthetic purposes.

7.2 Adding to the lexicon of post-production

The reality of operating as a creative sound practitioner on television or feature film production is that you rarely start with a figurative blank canvas; and from the point at which the post-production sound department will be brought on to a project, often a temporary soundtrack will already have been born. Certainly, significant decisions that can impact on the overall sound design might already have been made: e.g. the pace of a scene's cutting; takes are chosen that suit visual objectives better than aural; temporary sound effects may be put in place by the picture editor; and similarly, the use of 'temp' music in the absence of the commissioned score. All of these are things that can impact on the final mix and limit the creative options of the Re-recording Mixer.

In short, it is an inescapable truth that the vast majority of moving picture soundtracks, such as those for commercials, television dramas or documentaries (as well as feature films, but to a somewhat lesser extent), actually have their genesis not under the control of a Sound Designer, but that of a picture editor, maybe even before the Supervising Sound Editor has been engaged, and usually before the individual sound disciplines of dialogue and ADR editors, sound effects editors, Foley editors and music editors are brought on to start work.

When it is the case that the sound department are working with the *fait accompli* of the picture department's supplied audio, the creative space left is predominantly within the mix: the balancing of the supplied sound elements, which is driven by the emotional interpretation and physical crafting of the Re-recording Mixer.

Which is why presenting the final mix to a Director who is more familiar with several weeks of listening to the picture editor's (arguably) less appropriate balance of sound effects and music, can sometimes come as a rather uncomfortable, or even unwelcome, revelation.

But where such discomfort or disharmony does occur between the Re-recording Mixer and the Director, perhaps through a disagreement on a finer point of the plot, or a misunderstood directive, reference to the Four Sound Areas framework could go some way to help a Director or Picture Editor to convey more easily – and the

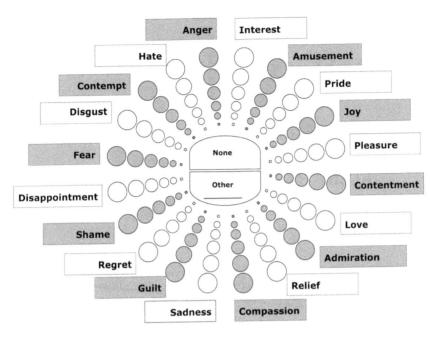

Figure 7.1 Monochrome Geneva Emotion Wheel

Re-recording Mixer to understand more readily – the emotional direction that was in mind when a scene was written, shot, tracklayed or cut.

Because the trade shorthand or technical terms used routinely by the sound department are not necessarily the most effective way for non-sound professionals to communicate and share their artistic aspiration, a wider understanding and adoption of the Four Sound Areas framework could allow Directors or Editors to ask the Re-recording Mixer at the mix (or agree with the Sound Designer at the spotting session) if the balance of a particular scene or event may be, for instance, steered more towards the Abstract sound area rather than the Narrative sound area, or vice-versa.

A shorthand to facilitate this, however, would need to be sought.

With these considerations in mind, I thought that a final element of the Four Sound Areas framework was required – a simple visual aid that could be used as a reference alongside the conceptual Four Sound Areas; and my proposed viewable representations are based on the Geneva Emotion Wheel (GEW) self-reporting tool (see Scherer (2005); Scherer, Fontaine, Sacharin, & Soriano (2013)). It is shown in its original form in Figure 7.1.

The GEW displays a spectrum of 20 emotional feelings, with an intensity ranging between 1 (lowest) and 5 (highest). A basic template of this sort would allow the Director, and the Sound Designer or Re-recording Mixer to refer to the emotional descriptors set around the periphery of the GEW, and use them as a way to

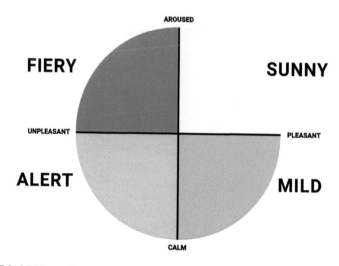

Figure 7.2 GEW emotion sectors categorized by colour, adapted for use as a Four Sound
Areas 'Mix Disc', with a summary word for each quadrant: Yellow = 'Sunny',
Green = 'Mild', Blue = 'Alert', Red = 'Fiery'

discuss, focus and map the kind of emotional experience intended by the Director,
or proposed by the Sound Designer, scene by scene.

Moving forward from the basic GEW template of Figure 7.1, I arranged coloured
quadrants (shown in Figure 7.2) to be superimposed on top of the standard monochrome Geneva Emotion Wheel. This I have called the Four Sound Areas *Mix Disc*.

It was inspired by the use of colours on the GEW in Sacharin, Schlegel and
Scherer (2012); as well as my desire to adapt and include the spirit of a classification for four basic personality types, that was first proposed by psychologist
William Moulton Marston in his book *Emotions of Normal People* (1928).[1] I felt
that later applications of Marston's original work unwittingly offer an interesting
relevance to the considerations that are applied at the design of moving picture
soundtracks.

By the 1970s, Marston's pioneering work had been greatly developed, refined
and adapted by people such as Walter Vernon Clarke (1956) for use by employment recruitment specialists under the generic term, DiSC profiling.[2]

Versions of this personality investigation process have stood the test of time
remarkably well, and derivatives of Clarke's original self-reporting tests are still
in commercial use today all over the world, promoted by companies like Wiley in
the United States, and the Swedish human behaviour analytics company, Ensize,
throughout Europe.

Each of the colours of the *Mix Disc* quadrants in Figure 7.2 represents a 'zone'
containing five of the GEW emotional descriptors, which I have summarized in a
single word: Yellow is represented as the 'Sunny' sector of emotions; Green by
the word 'Mild', Blue by the word 'Alert' and the Red zone by 'Fiery'.

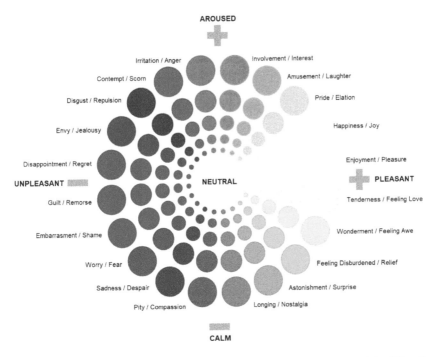

Figure 7.3 GEW adapted by this study for use as a Four Sound Areas 'Mix Wheel', categorized by colour and showing all 20 emotion descriptors

A vertical axis moves from a state of 'Calm', through neutral at the origin, to 'Aroused'; whilst a horizontal axis moves from 'Unpleasant', through neutral, to 'Pleasant' feelings.

An emotion that is considered unpleasant (negative) or pleasant (positive) is represented by which half (hemisphere) of the GEW the emotion descriptor is situated on: the left half representing negative emotions, the right half positive emotions (with respect to the 'x-axis').

The bottom half of the hemisphere contains gradations of 'Calm' towards neutral, whilst the top semi-circle represents a progression from neutral towards 'Aroused' (with respect to the 'Y-axis'). The degree to which an emotion is felt remains the same as on the original GEW scale – between 1 (lowest) and 5 (highest).

These coloured subdivisions have developmental relevance to the initial diagrams used in Figure 5.2 (the sound 'quadrant' used for *The Craftsman*) and Figure 5.7 (itself a development of the sound 'compass' used for *Here and Now*) inasmuch as their intended purpose was to be a simple visual aid for, e.g. the Sound Designer and Re-recording Mixer to visualize and communicate the Director's intended emotional experience for a particular scene; or vice-versa.

The second adaption of the GEW model, shown in Figure 7.3, I have called the Four Sound Areas *Mix Wheel*, and again, it is proposed for use by Directors and Sound Designers in spotting sessions. Easily reproduced and printed, it is suggested that this could also play a useful role either before, during or after the spotting session; this version being used to provide the finer detail for emotional intent. Separate sheets for different scenes could be provided as a record for later reference, along with any notes and accompanying paperwork, at the mixing stage.

The differences between the *Mix Disc* and the *Mix Wheel* are easy to see: in the *Mix Wheel* (Figure 7.3) the colours graduate and almost merge as the emotional descriptors move from one to another around the wheel – just as individual sounds blend in a soundtrack, through the act of mixing; whilst in contrast, the edges of the coloured quadrants in the *Mix Disc* (Figure 7.2) are more precisely differentiated.

It is suggested that the *Mix Disc* template can be used as a way to quickly set the initial emotional positioning of a scene; with the fine-tuning of the intended emotions being able to be recorded using the more detailed *Mix Wheel* template. Mixing sheets may be produced with the annotated *Mix Disc* on one side and the *Mix Wheel* on the other.

7.3 Examples of the Four Sound Areas Mix Wheel in use

Noting the key scenes from *The Craftsman* (as discussed in Chapter 5 and shown in Figure 5.2) and recording them on a *Mix Disc* 'summary sheet' produces the general emotional placement of all the key scenes, as shown in Figure 7.4.

A more detailed guide, utilizing the *Mix Wheel* (e.g. annotated scene-by-scene, following discussions with the Director at a spotting session), could produce sheets similar to those in Figures 7.5–7.12.

In this instance, I have created the *Mix Wheel* sheets for *Here and Now*, using the scenes discussed in Chapter 5 (and illustrated in Figure 5.7 and Table 5.1), showing the intended emotional reference points.

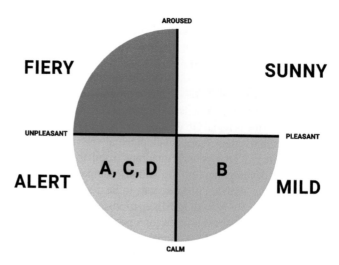

Figure 7.4 Mix Disc for *The Craftsman*, providing an overview of complete film

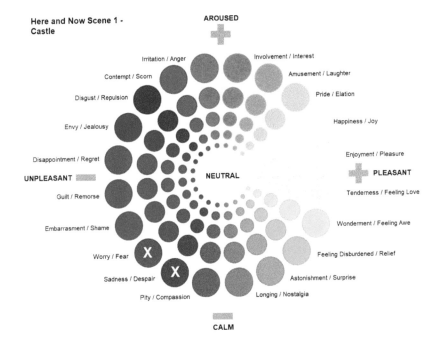

Figure 7.5 Here and Now Mix Wheel - Castle scene

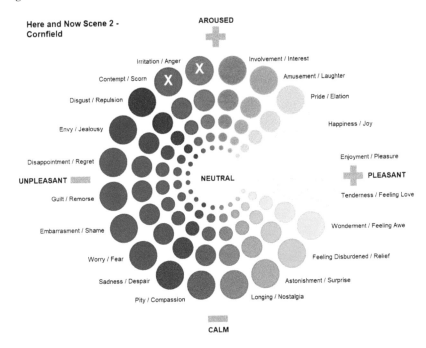

Figure 7.6 Here and Now Mix Wheel - Cornfield scene

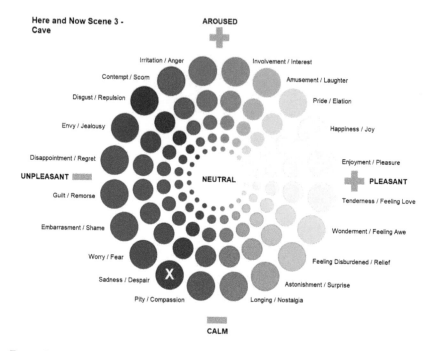

Figure 7.7 Here and Now Mix Wheel - Cave scene

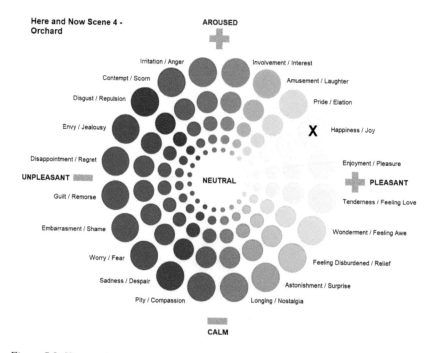

Figure 7.8 Here and Now Mix Wheel - Orchard scene

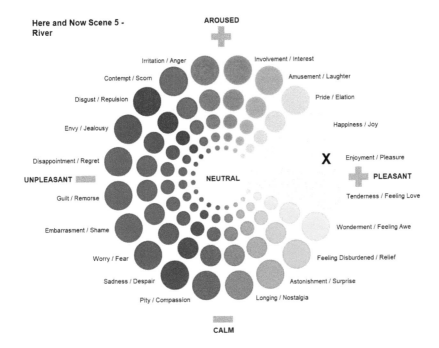

Figure 7.9 Here and Now Mix Wheel - River scene

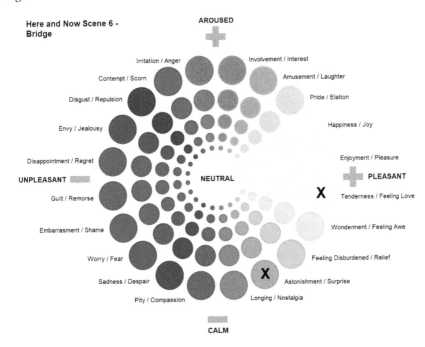

Figure 7.10 Here and Now Mix Wheel - Bridge scene

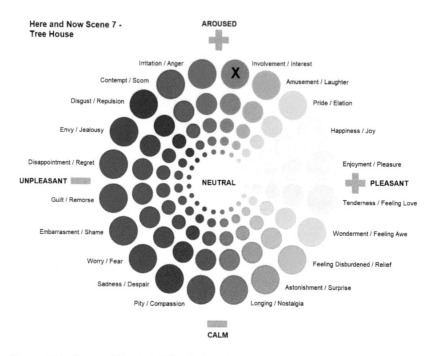

Figure 7.11 Here and Now Mix Wheel - Tree house scene

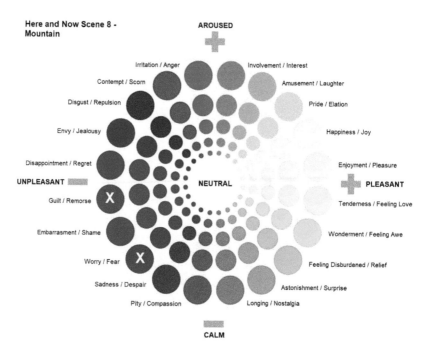

Figure 7.12 Here and Now Mix Wheel - Mountain scene

7.4 The Four Sound Areas: a beginning and an end

The use of the *Mix Disc* and the *Mix Wheel* neatly brackets the sound design process: at the initial stage, they are useful tools for recording soundtrack concepts with the Director, and represent an opportunity for the Sound Designer to begin structuring ideas, mapping the intended emotional journey and creating a sense of order towards the sound design aspect of the audio post-production process.

And then, at the final mix stage, the *Mix Disc* and *Mix Wheel* sheets can serve as useful tools to enable the Re-recording Mixer to look back at the initial emotional intentions for the soundtrack.

7.5.1 Discussion points

- Audio post-production, of which Sound Design forms an essential part, conventionally begins after the picture editing stage is completed. How might the use of the Four Sound Areas, utilizing the Mix Disc and Mix Wheel tools, aid the picture editor and Director to consider the soundtrack at a much earlier stage?
- What advantages or disadvantages might this create?

7.5.2 Practical exercises

- Choose a selection of your favourite film scenes and complete both a Mix Disc summary and a more detailed Mix wheel.

Notes

1 Not only a leading psychologist of his time, with his wife Elizabeth Holloway, Marston invented the prototype of the lie-detector test. He was also a comic book writer, creating the character of Wonder Woman in 1941, partly in tribute to his polyamorous partner Olive Byrne. He entered the Comic Book Hall of Fame in 2006.
2 The *DiSC* Model of Behaviour has its origins in William Moulton Marston's work *Emotions of Normal People* (1928). Marston proposed that the behavioural expression of emotions could be categorised into four primary types: Dominance, Influence, Submission and Compliance. In 1948 it was developed into an assessment instrument by industrial psychologist Walter Vernon Clarke (1956) and in 1958 Walter Clarke Associates launched a recruitment tool called *Self* DiSC*ription*. Based on this work, the process of DiSC profiling has remained a staple tool of recruitment, used to judge an applicant's employment suitability.

8 Closing thoughts

Introduction

In this chapter I look at further areas of possible research and summarize what possibilities exist for filmmakers who are ready to embrace sound design more fully.

8.1 Further considerations

Happily, I believe that there is still ample scope for academics, students and practitioners to undertake further research within the field of sound and emotions – especially in the kind of work that is directly related to the sound design and mixing of soundtracks for moving pictures; and in particular, in such work that bridges the gap between academic understanding and practitioner methodology.

It seems to me that there is distinct value in furthering this knowledge – whilst developing a greater practical understanding – of how sound design may successfully evoke intended emotions within a listening-viewer, through the soundtrack.

Other questions that strike me as worthy of further investigation include:

- Can a 'Universal Law' for emotive moving picture sound design be established that works for a variety of media, spanning auditioning platforms such as movie theatre, home theatre, television, on-line, hand-held and gaming?
- How important is dialogue delivery to the effectiveness of conveying emotion?
- How can editing the rhythm and pace of dialogue influence a listening-viewer's emotions?
- What natural, non-human sounds, either as background atmospheres or as foreground spot effects, elicit clear, repeatable emotional responses?
- How significant is the configuration of the auditioning channel to emotional response, i.e. monophonic, stereophonic or surround sound presentation? Can a Monophonic soundtrack ever be as emotive as a Dolby Atmos multi-channel soundtrack?
- Does the presentation medium pre-dispose the listening-viewer to respond in a certain way? (e.g. cinema versus home). How close an emotional experience

can a home theatre or hand-held device come to that of the carefully designed movie theatre soundtrack?

- Can re-purposed content – such as movies used for in-flight entertainment, or television drama streamed by commuters contending with inherently high background noise – evoke emotions consistent with other auditioning environments and platforms, because subtle elements are difficult to hear in the mix?
- Are a listening-viewer's emotional responses noticeably different between real-life moving pictures and artificial images? Should certain sound design elements be more exaggerated for animation, given that the audience know that what they are seeing is not real? Do things alter in the ability of sound to evoke audience emotions when using different types of animation: such as stop motion, 2D or 3D animation?
- How suitable and how easily could the Four Sound Areas integrate into Augmented Reality (AR) and Virtual Reality (VR) applications?
- How could qualitative data best be gathered from users and audiences to include their feedback and comments?

8.2 Closing thoughts

Ingmar Bergman's quote on the front cover of Tarkovsky's autobiography *Sculpting in Time*, described him as 'the most important director of our time'; and both Bergman and Tarkovsky directed their movies with passion, and from an *auteur* perspective: their artistic vision is clear to see as the driving force for their films. Both were considered attentive to their soundtracks, Bergman arguably the more so; but over 30 years ago, Tarkovsky, a Russian émigré filmmaker exiled in Paris, hinted at the untapped potential within the soundtrack to convey artistic truth, when he wrote:

> I have a feeling that there must be other ways of working with sound, ways which would allow one to be more accurate, more true to the inner world which we try to reproduce on screen; not just the author's inner world, but what lies within the world itself, what is essential to it and does not depend on us. (Tarkovsky, 1987, p.159)

Yet from my experience as a moving picture sound professional, contributing to productions that range from mainstream US studios (e.g. *Lincoln, New York, I Love You, Pirates of the Caribbean*) to more modest UK 'indie' movies (e.g. *Kingsman: The Golden Circle, Finding Fatimah, Scott and Sid*) and micro-budget films (e.g. *Mountain Biking: The Untold British Story, Last Shop Standing, Money Kills*) I frequently observe that amongst non-sound practitioners working within the film industry, there is less of an acceptance than that shown in academic circles, that sound is truly able to enhance the ability of pictures to tell a story.

Instead, it would often seem that sound is considered as the fall-back option for when the pictures don't quite tell the story, or when a scene needs repairing;

reinforcing the notion that first comes the picture, then comes the sound: a patronizing statement for both sound and vision that does little to move either set of craft skills forward, or develop the medium.

Bresson is clear on the egalitarianism between sound and picture:

> When a sound can replace an image, cut the image, or neutralize it. The ear goes more towards the within, the eye towards the outer. (Bresson, 1977, p.28)

And yet by speaking out and championing the importance of sound, it inadvertently manages to suggest a continuation of sound's subservient status to pictures, particularly to those who continually choose to not listen. Writer, Producer and Director George Lucas comments on this equipoise:

> I feel that sound is half the experience. Filmmakers should focus on making sure the soundtracks are really the best they can possibly be because in terms of an investment, sound is where you get the most bang for your buck. (Blake, 2004)

However, modern, and evolving film production methods, contemporary post-production technology, audience taste and consumer hardware mean that an existing outmoded and hierarchical approach towards sound could be reaching an overdue extinction. Sound Editors are now able to work much more collaboratively, and earlier, in the post-production process with picture editors; allowing the sound and picture departments to start work together on the first cut, and then on throughout the subsequent post-production stages.

Indeed, with the introduction of version 15 of their holistic, sound-and-picture post-production application in mid-2018, Black Magic Design's DaVinci Resolve became the first fully integrated sound and picture solution to cover media ingest on location, picture editing, colour correction, visual effects and audio post-production, sharing a common media pool and timeline.

In post-production, this enables picture editors, colourists, VFX artists and sound editors to work simultaneously on the same project, without the need to export media, file exchange and re-import, and this type of collaborative working could point the way towards a much more efficient way forward. It is, for instance, feasible for the spotting session to take place much earlier in the picture editing workflow (e.g. at the time of the 'rough cut') so that a Picture Editor's presence at the spotting session, alongside the Director and Sound Designer, could deliver a commonality of purpose, and at least in terms of intended emotion.

To that end, this new kind of flexible workflow could not only be hugely influential in bringing sound and picture departments closer together – and earlier – in creative collaboration; but it could also be the point at which a new sound mixing framework for enhanced emotive sound design within contemporary moving picture audio production and post-production could also be timely.

Regarding championing the role of sound in filmmaking, Sound Designer Gary Rydstrom suggests that:

> If we do our jobs well and throw in a little evangelizing, we can make sound as important a part of filmmaking as it should be. (Kenny, 2004)

The desire of both the film and television industry is to deliver increasingly enhanced aural experiences, and the appetite for this from consumers is clearly demonstrated by the plethora of audio platforms, equipment and software that is manufactured specifically for the originating of immersive and emotive sound in audio production and post-production; and for the replaying of these increasingly sophisticated soundtracks within film theatres, and on home cinema set-ups.

For consumers, the choice and understanding of seemingly ever-evolving sound technology can be bewildering. Denison (2017) notes:

> sound, for many, remains a confusing technology. Though most understand the concept of using multiple speakers for theatre-like sound, many don't understand the difference between all the different formats. There's 5.1, 6.1, 7.1, 9.2, Pro Logic IIx, Pro Logic IIz, Dolby DSX and more. It's a lot to wrap your head around. (Denison, 2017)

And so, as film sound playback technology has evolved towards the immersive experience audiences can now hear at home as well as in modern movie theatres; the weight of responsibility for the soundtrack to support visual content through dialogue and music alone has been increasingly shared by other soundtrack elements. Today's rich soundtracks are constituted with very much more than the traditional bastions of dialogue and music; and the purpose of this study is to propose and demonstrate that the Four Sound Areas framework can encourage storytelling to arise more easily, and develop more naturally, out of the soundtrack.

Meanwhile, multiple Oscar-winning Director Steven Spielberg understands and voices what Sound Designers and Re-recording Mixers passionately feel:

> The eye sees better when the sound is great.
> (Barsam and Monahan, 2009, p.368)

The commitment behind the research and creative practice that resulted in this book, is an attempt to serve both the rigours of academic argument and to also provide a useful approach for practitioners of all disciplines engaged with moving picture soundtracks (e.g. picture editors, Directors, as well as sound editors, Sound Designers and Re-recording Mixers), highlighting how and what the creative input from the Sound Designer and Re-recording Mixer can add to an audience's emotional experience of a film, even at the final stage of its production process; and also to help promote a wider understanding that the thoughtful use of sound can bring a greater impact to any moving picture production, through the often untapped potential that exists connate when filmmakers actually take the time and trouble to consider the conclusion of this study: that audience emotions are more easily, and most effectively, accessed by the thoughtful use of emotive sound by an enlightened Sound Designer, which is purposefully expounded through the sympathetic mixing of a skilful Re-recording Mixer.

Ultimately, this book aspires to be a catalyst for change, igniting interest, informing practice, and inspiring a new generation of filmmakers to earn their own place in the pantheon of great film Director/Sound Designer collaborations, notably including:

Alan Crosland with Nathan Levinson (*Don Juan, The Jazz Singer*); Akira Kurosawa with Ichirô Minawa (*Seven Samurai, Throne of Blood*); Ingmar Bergman with Evald Andersson (*The Seventh Seal, Persona*); David Lean with Winston Ryder (*Lawrence of Arabia, Dr. Zhivago*); Stanley Kubrick with Bill Rowe (*A Clockwork Orange, Barry Lyndon*); Francis Ford Coppola with Walter Murch (*The Conversation, The Godfather Part II*); Andrei Tarkovsky with Semyon Litvinov (*Solyaris, The Mirror*); Robert Altman with Richard Portman (*Nashville, 3 Women*); Steven Spielberg with Gary Rydstrom (*Saving Private Ryan, War Horse*); George Lucas with Ben Burtt (*Star Wars Ep. IV, Star Wars Ep. V*); Steven Soderbergh with Larry Blake (*Ocean's 11, Solaris*); David Lynch with Alan Splet (*Eraserhead, The Elephant Man*); Joel and Ethan Coen with Skip Lievsay; (*Barton Fink, The Big Lebowski*); Danny Boyle with Glen Freemantle (*28 Days Later, Slumdog Millionaire*); Christopher Nolan with Richard King (*Inception, Dunkirk*).

Ultimately, the call to action from this book is to recognize that there is an opportunity for you, too, to appear in the audio Akashic records, if you choose to understand, adopt and believe in one simple notion: *only by fully embracing the soundtrack will you truly create moving pictures.*

9 Appendices

Appendix 1

9.1 Release dates of the feature films utilized in this study

The Craftsman (Dir. L. Murphy) – premiered 10-05-12, Stafford, UK.
Here and Now (Dir. L. Turner) – premiered 11-03-14, Malmö, Sweden.
Finding Fatimah (Dir. O. Arshad) – premiered 06-04-17, London, UK.

Appendix 2

9.2.1 **Here and Now***: other significant sound design details*

[01:03:17:03–01:07:20:01] Predominantly utilizing the *Narrative* and *Abstract* sound areas, the atmosphere tracks laid-in here first establish and then illustrate the contrast between the city (seen from [01:00:20:15] and heard from [01:0048:00 to 01:01:20:00]) and the country – through, at first, prominent morning birdsong, which is then joined by the sound of an outside, distant tractor working the fields as Grace moves through the house, to a burst of nature as Grace steps out of the cottage for the first time (01:04:56:06–01:07:20:01). However, the only time we see the working machinery hinted at in this sequence is at 01:43:40:20–01:43:46:11, when a tractor pulling a trailer across a field can be seen at some distance away.

[01:10:06:00–01:10:46:18] Predominantly utilizing the *Temporal* sound area, the music track that replaced this temporary version's jolly tune originally cued so that a *glissando* down occurred each time Say rode his bicycle away from Grace, as if she were disappointed by him leaving; however, I felt that that did not accord with her dismissive attitude towards him, seen immediately before the riding sequence. So, the replacement music was re-cued so that each time Grace rode up and alongside Say, she would bring a 'down-beat' mood-change with her, reflected in the music. Say would again ride away, the music would lighten, only for Grace to catch up and the mood of the music to darken once more.

[01:19:17:14–01:19:26:15] Predominantly utilizing the *Abstract* sound area, the emptiness of the village green, its inherent bleakness supplemented by the sinister squeak of swings moving in the wind, contrasts with Grace sunbathing [01:18:45:04–01:19:17:14] to the accompaniment of rather more reassuring birdsong.

[01:58:40:00–02:03:28:00] The emotional 'low' – or 'tear-jerker' – point of the film was by design the point at which Say's mother Poppy breaks down at the party and tells Lucy that her husband – Say's father – died on the mountain; and that Say has blamed himself for this. This was mixed so that the *Narrative* dialogue is unimpeded by virtue of subtly dropping the level of the *Abstract* crowd sounds in the front speakers during this speech, with an emotively sentimental, guitar-based song entering the timeline at [02:01:57:00] (*Temporal* music and *Narrative* words), whilst we hear and see Poppy crying and being held by Lucy. I painstakingly edited the originally supplied song to create an extra eight bars of instrumental introduction to better sit the music against the montage pictures (*Temporal* music sitting with the *Abstract* sounds of children playing), ahead of where we eventually see the singer/guitarist in vision. The song also intentionally transitions during this sequence from being non-diegetic to diegetic, the longer instrumental introduction better serving the non-diegetic, reflective purpose of the song.

Appendix 3

9.3.1 Background to the XX Commonwealth Games in Glasgow

Over the 12 days of competition between July 23rd and August 3rd, 2014, SVGTV – the joint venture between UK sports producer Sunset and Vine and Australia's Global TV – acted as the Host Broadcaster and provided Rights Holding Broadcasters around the world with over 1,200 hours of live coverage from the 17 sports that made up the twentieth gathering of the Commonwealth Games.

In the UK, the audience for the BBC's relaying of the 2014 Commonwealth Games was estimated to be around 6.8 million viewers per day (BBC, 2014); a far cry from the total of 51.9 million people, or 90% of the population, who tuned-in to the 2012 London Olympic games on BBC Television (Plunkett, 2014). However, the television audience worldwide for the Glasgow games was a much healthier figure, with an estimated one billion viewers spread mainly across the 71 nations of the British Commonwealth (Kellaway, 2014).

In contrast to single camera shooting on drama, where one or two boom microphones may be deployed, and actors might each wear a personal microphone – a low microphone count – the number of microphones that require mixing together is much higher for sports Outside Broadcasts; employing many microphones in fixed positions relevant to the action.

The complexity of capturing and transmitting live O.B. sound from its source is often in stark contrast to the relative simplicity of either recording sound for filmed drama on location or editing and mixing in a post-production studio. An O.B. can often involve the audio equivalent of carrying out both the production and post-production stages of a film simultaneously.

The finals for all the boxing categories, male and female, took place in the Hydro at the SECC precinct. With a capacity of 11,000, this was a substantially larger venue than that of the heats' Hall 4a, which seated a smaller audience of 3,000.

I was afforded creative freedom by the Production team to achieve the required sound coverage, but within certain unavoidable constraints; e.g. suspended hypercardioid microphones over the ring were ruled out of bounds as they would impinge on the clean overhead shot of the remote-controlled Camera 7; and the desired hypercardioid microphones looking down into the ring, mounted to the lighting truss suspended above the ring, could not be physically attached in time, given the logistics of an overnight de-rig of Gymnastics and the re-rigging for Boxing. Compromise is an important element of microphone placement on Outside Broadcasts.

The audio outputs for transmission purposes were rigidly defined: a full stereo effects mix without Commentary (called 'International Sound', designated 'TVIS'); a full stereo effects mix including Host Broadcaster commentary (called 'English Language Commentary for Television with International Sound', designated 'ECTVIS'); a stereo effects mix for radio without commentary, which was also sent minus any audio that directly related to pictures, such as the feeds from the close-up camera microphones, the microphones placed in the boxers' corners and the EVS replay machines (called 'Radio International Sound', designated 'RAIS'); and a clean Commentary feed. There were also mono down-mixes of TVIS and RAIS derived within, and sent out from, the mixing desk.

Appendices 9.3.2 and 9.3.3 show the desk configuration and layout taken from the heats, which was transferred to the finals as a starting point for configuring the Calrec Apollo mixing console. The initially spare main output group (Group 5), was used in the finals to control a PA feed to the RAIS circuit, solely for the live bagpiper and recorded music of the medal ceremonies; the PA being a feed that would not otherwise have been included in the RAIS outgoing signal.

9.3.2 Planning sheet 1 for Commonwealth Games Boxing heats/finals

Neil Hillman - Audio console configuration - Calrec Apollo

XX Commonwealth Games Glasgow 2014 - Boxing rig

Main Output Groups (and Destination)

1 - Int FX
2 - Pgm FX
3 - Radio FX
4 - Hi-Levels
5 - (Spare)
6 - Clean Commentary
7 - TVIS stereo
8 - ECTVIS stereo
9 - RAIS stereo
10 - TVIS mono (fold down)
11 - ECTVIS mono (fold down)
12 - RAIS mono (fold down)

Embedded Outputs

A1: L+R - TVIS (TV International Sound) stereo
A2: L - Directors talkback
A2: R - Clean commentary
A3: L+R - ECTVIS (English Commentary for TV mixed with International Sound) stereo
A4: L+R - RAIS (Radio International Sound) stereo

Fader	Layer	Source	VCA group	Group	Output Group
1	1	Crowd (wide) stereo (2 x 416)	1	1,2,3	See sheet 2
2	1	Crowd (centre) MS	1	1,2	
3	1	Ring near - 816	2	1,2,3	
4	1	Ring far - 816	2	1,2,3	
5	1	Red corner plant - ECM		1,2	
6	1	Blue corner plant - ECM	2	1,2	
8	1	Cam 3 - 416		1, 2	
9	1	Cam 4 - 416		1, 2	
10	1	Cam 5 (RF) - 416		1, 2	
12	1	Commentary		6, 8	
13	1	Crowd Group fader - (VCA 1)			
15	1	Ring FX Group Fader - (VCA 2)			
17	1	Flash Position - RE 50		1, 2	
19	1	PA high level		4	
20	1	PA VOX		4	
22	1	EVS - Red		4	
23	1	EVS - Blue		4	
24	1	EVS - Gold		4	

Figure 9.1a Commonwealth Games planning sheet 1

9.3.3 Planning sheet 2 for Commonwealth Games Boxing heats/finals

Neil Hillman - Audio console configuration - Calrec Apollo

XX Commonwealth Games Glasgow 2014 - Boxing rig

Fader	Layer	Source / Destination
25	2	Group 1 - FX International
26	2	Group 2 - FX Programme
27	2	Group 3 - FX Radio
28	2	Group 4 - HI Levels
29	2	Group 5 - (Not used)
30	2	Group 6 - Clean Commentary - mono
31	2	Group 7 - TVIS - stereo
32	2	Group 8 - ECTVIS - stereo
33	2	Group 9 - RAIS - stereo
34	2	Group 10 - TVIS - mono
35	2	Group 11 - ECTVIS - mono
36	2	Group 12 - RAIS - mono
37	2	Aux 1 - PA Vox to commentators box [marked PA on box]
39	2	-18dB stereo tone source, routable to all outputs

Main Output Groups (and Destination)

1 - Int FX
2 - Pgm FX
3 - Radio FX
4 - Hi-Levels
5 - (Spare)
6 - Clean Commentary
7 - TVIS stereo
8 - ECTVIS stereo
9 - RAIS stereo
10 - TVIS mono (fold down)
11 - ECTVIS mono (fold down)
12 - RAIS mono (fold down)

Embedded Outputs

A1: L+R - TVIS (TV International Sound) stereo
A2: L - Directors talkback
A2: R - Clean commentary
A3: L+R - ECTVIS (English Commentary for TV mixed with International Sound) stereo
A4: L+R - RAIS (Radio International Sound) stereo

Figure 9.1b Commonwealth Games planning sheet 2

Appendix 4

9.4.1 Finding Fatimah: *background, collaboration and sound design*

Finding Fatimah is a low-budget[1] feature film made by British Muslim Television (BMTV), a free-to-air education and entertainment channel broadcasting 24 hours a day in the UK on the Sky TV platform. It was written and directed by Oz Arshad and released in British cinemas by a major distributor, Icon, in April 2017. Its Producer was Sol Harris.

It introduced two new actors in the lead roles (Danny Ashok as Shahid and Asmara Gabrielle as Fatimah) and featured cameo roles from established actors such as Nina Wadia, Guz Khan, Denise Welch and Dave Spikey.

The film is about a Muslim man (Shahid) who struggles to find love in the British Asian community due to the stigma of his divorce several years prior; a theme that Arshad decided to write about from his own experience of being Muslim and divorced. There were certain restrictions applied on the script by the financiers, and given that the movie's genre was romantic-comedy ('rom-com') it meant that careful attention was paid throughout the production process to ensure that the film appealed to both conservative and liberal Muslims, as well as a mainstream audience.

In an interview given to promote the film, Director Arshad said:

> The universal theme of this movie is finding love and these two characters just happen to be Muslim. (Kayani, 2017)

The film was shot on location in Manchester and West Yorkshire, in segments, between July and October 2016, with some extra filming taking place in January 2017.

Whilst the film was an unmitigated joy to be a part of, and received plaudits for its production values, it received a mixed reception at its launch. It was certainly well received by its target audience of British–Asians, who made up a significant number of the cinema-going audience, and it exceeded its initial box office expectations; but ultimately it failed to capitalize as a 'cross-over' film that appealed to both British–Asian and British–White audiences, in the way that other British–Asian romantic comedies such as *East is East* (1999) or *Bend It Like Beckham* (2002) had previously done.

The magazine *Total Film* gave the movie a creditable 3 out of 5 stars (Coleman, 2017) whilst *The National Student* went one better (Storr, 2017). Meanwhile, MaryAnn Johanson writing in *Flick Filosopher* commented on its production values:

> Arshad makes his ridiculously low-budget film look far more expensive: I would never have guessed that Finding Fatimah was made for under half a million pounds. That's just nuts. (Johanson, 2017)

But the *Guardian* newspaper film critic Peter Bradshaw was rather less effusive, describing it as a faintly depressing comedy (Bradshaw, 2017).

9.4.2 Plot

A young, divorced Muslim man (Shahid) is looking for love, but try as he might, he cannot find a suitable partner to become his wife. Now approaching his thirtieth birthday, and with his computer printer business on the verge of collapse, he is conscious of his loneliness and the stigma attached to his divorced status. So, he signs up to a Muslim dating site and with the unbidden help of a work colleague who selects likely candidates on his behalf, he is computer-matched to several unsuitable women. One of these, however, (Fatimah) noticeably stands out from the rest.

Fatimah is an independent and ambitious medical Doctor of a similar age, and from a strict Muslim family. Whilst her father is keen to arrange for Fatimah his choice of husband, she meets Shahid and they fall in love. But Shahid struggles to tell her about his previous marriage; and Fatimah has her own secrets, too: she has an anger problem[2] and amongst the list of requirements for any future husband, there is one that Shahid can never meet. And along the way, Shahid is progressing through the rounds of a television talent show as a stand-up comedian, delivering his own material about Muslim families and his lack of luck in love.

9.4.3 Audio post-production and collaboration

Picture editing and audio post-production began in January 2017, and the dialogue, ADR, music editing and pre-mixing was carried out by me at The Audio Suite, Birmingham. The task of backgrounds, sound effects and Foley creation was undertaken by my friend and colleague Anna Sulley,[3] who acted as the Supervising Foley and Sound Effects Editor; with Sue Malpas[4] sharing some of the Foley recording duties in her Essex studio.

Final mixing for cinema, TV and online distribution was carried out in the Powell theatre[5] at Pinewood studios, Buckinghamshire, by my long-standing friend and Re-recording colleague, Pip Norton.[6]

Initial sound ideas and concepts came from discussions with the Director at a spotting session held at The Audio Suite, led by me, with sound effects and Foley supervisor Anna present. At this meeting several decisions were made, such as the 'feel' for each scene, the comic timing and placement of sound effects and the sound of the major location ambiences. These atmospheres also had the time of day and the socio-economic positioning of their environment taken into account. Many original spot effects were created by Anna for specific items such as phone message alerts from the dating site's 'phone app', and audio stings for the graphics of the on-screen television show, *Muslims With Talent*; with abstract tonal elements and non-diegetic sound effects synthesized and layered to enable them to work seamlessly in the mix alongside the score.

Notwithstanding that the film was final mixed on one of the UK's largest film consoles, in a large mixing theatre, as the film was pre-mixed at The Audio Suite by me bringing together all the dialogue, sound effects and music elements 'in the box' (my usual method of working), it was decided that this would continue at Pinewood, for the Theatrical Mix; with Pip Norton taking care of the physical 'mechanics' of mixing, rather than there being a traditional division of labour at the mixing console: where one pair of hands and ears balances the dialogue and music against each other, whilst a second pair of hands and ears mixes the backgrounds and effects, and drives the machinery.

It is also worth noting the way that pre-mixes and media are transferred to the final mix stage: reel-by-reel pre-mixes are rarely 'printed' in the traditional way anymore, i.e. rendered as single clip mix stems to be mixed against each other – instead, by sound editors working 'in the box', it is increasingly customary for each and every individual clip to be levelled ahead of the final mix; so that by importing the AAF for each reel of a project into a mixing theatre's workstation, and then playing back the tracks with the console and DAW faders at unity, *my* intended pre-mix of all the elements is presented. All the sound clips on the timeline are replaying at my intended levels, without the initial need for any fader movement; this 'semi-temp/semi-final mix' is the starting point for the final mix. (However, as any Re-recording Mixer will tell you, this is only a *starting* point; there will be myriad adjustments to these levels throughout the course of the final mixing stage.)

This method presents today's Re-recording Mixer with a busier final mix timeline to balance than perhaps previous generations faced; but it ensures easy access to all of the individual clips that make up the 'pre-mix', not just the recorded ('printed') or rendered stems of, e.g. dialogue, effects, atmospheres and music, should the Mixer need to adjust or replace an item.

It also highlights just how painstaking a tracklay is for the individual sound editors, needing to carefully level *every* clip, but it makes for a precise final mix. (The other point worth making is that I really value the input Pip brings to a final mix and therefore I always want her to have ready access to any component part of the soundtrack.)

As far as room reverbs are concerned on dialogue, whilst I always record into the timeline my chosen room simulation at the pre-mix stage, I keep the clean dialogue clips and their associated reverb clips on separate, but adjacent tracks; so that if at the final mix they need replacing for something different or more suitable, it is an easy thing to achieve within the project timeline.

With us working in this way, I was able to either carefully listen whilst sitting alongside Pip at the mixing desk, without any of the distractions of operating the console; or I could sit with the Director at the rear of the theatre, appraising the fuller experience of the film. And so together, Pip and I arrived at the balance and emphasis desired for each scene by discussion, re-winding and re-balancing as required.

In practice, this proved to be a highly efficient process, not least of all due to the exceptional skill, speed and dexterity of Pip as a Re-recording Mixer; the quality of the effects, backgrounds and Foley provided by Anna Sulley; the three of us having an understanding of each other and an established working relationship; a similar mindset and significant respect for each other's points of view; and through the discussions that had taken place, and the mixing notes that had been exchanged, ahead of the pre-mixes arriving with Pip for her to mix.

It meant that in my role of Supervising Sound Editor, overseeing all the soundtrack elements coming together from a team of people (albeit a small team of three), I had the luxury of being able to concentrate fully on the final product – a full, rich, theatrical-release soundtrack that conveyed the emotions we intended it to.

9.4.4 Four Sound Areas breakdown of selected scenes sound design

1 Fatimah finds herself subjected to unwanted attention in her gym.

[00:24:22–00:27:44] The scene opens with prominent music, initially working in the Narrative area to immediately suggest the high energy of a gym environment. Gradually, the music is blended with the background atmosphere of the gym's running machines and weight machines (in the Abstract sound area) by being treated to appear as if it is playing over the gym loudspeakers (the music moves from Narrative to Spatial). We see Fatimah and her friend Nayna together. The temporal element is provided by the added sound effects of the treadmill that Fatimah is running on (her footsteps added in Foley). Fatimah's heavier breathing was added in ADR as was Nayna's line, 'Babe, I feel like online isn't working' (for greater clarity than the original recording).

Nayna has Fatimah's mobile phone open on the dating app and is accessing her messages. We hear (and see superimposed on screen) the messages being 'swiped', and we hear (and see) the messages as they arrive (the message alert tones being in the Narrative sound area).

A new, non-diegetic music track starts, to signify Nayna noticing the arrival of an attractive man (a Narrative use of music) and it works to suggest that a desire for establishing sexual chemistry has been aroused in Nayna, right there in the gym. This music is quickly – and comedically – cut short, by the sound effect of a needle scratch across a vinyl record (a Narrative sound), coinciding with another gym user (Rocky) barging Nayna's vision of Adonis roughly out of the way, so that he instead may show himself off to the two women, and directly engage with Fatimah.

To ensure 'effortless intelligibility' of the dialogue between Rocky and Fatimah, the background gym music and effects are dropped to a very low level. Extra asides from Rocky were added in ADR as Fatimah draws him closer, and for greater clarity Rocky's final line to Fatimah was also replaced in ADR.

This scene would be described as operating in the Yellow (Sunny) quadrant of emotions, primarily (on display are the emotions of interest, amusement and pride); although Fatimah's display of anger, at the extreme of the Red quadrant (Fiery), directly adjoins the Yellow segment of emotions. (See Figures 7.2 and 7.3.)

2 Shahid and Fatimah fall in love at a charity pottery-painting event.

[00:35:39–00:39:49] This scene comprises of one of the essential elements in the formulaic structure of a rom-com: the point at which the two protagonists realize that they are in love and sentimental music overwhelms the sound effects for emotional intent.

Shahid arrives at an indoor charity event organized by Fatimah and the calm background hubbub of conversation (in the Abstract sound area) is used to paint the scene of a calm but busy environment. From the point of their initial greeting, the background conversation begins to diminish in level, leaving only Shahid and Fatimah's dialogue set against a slow and romantic, non-diegetic music track that is cued to begin softly, as Shahid delivers his first line to Fatimah. The music starts as a coy, halting piano figure, whilst we watch as Shahid and Fatimah's closeness becomes apparent to them, and us.

Increasingly, their lines of dialogue begin to swim intentionally in a syrup of sentimentality, provided by the harmonic progression of the music: woodwind instruments join the single piano notes, which have developed into exploratory chords (the music is in the Narrative sound area, telling us – in case we missed the signs – 'these guys are in love!'). As the music approaches some kind of crescendo, it is abruptly stopped as Shahid is addressed by a fellow attendee with an Arabic greeting; and he and Fatimah are brought out of their obvious reverie for each other (a reprise of the comedic musical interruption used in the Gym scene). This is a scene where the core intended emotions are simply defined: those of un-alloyed love and mutual admiration, aligning with the Green (Mild) quadrant. (See Figures 7.2 and 7.3.)

Fuller examples of romantic musical interludes can be found between [00:43:22–00:44:28] and [01:02:56–01:04:40]; with the obligatory rom-com 'heartache' musical montage at [01:10:41–01:13:13]. (Cued in the tracklay so that the line from Shahid's neighbouring diner 'Fix_My_Heart' ('I knew you weren't a good Muslim') carefully fitted between lines 4 and 5 of the lyrics in the song *The First Cut Is The Deepest* (Comp. Cat Stevens)).

3 *Shahid unexpectedly encounters his ex-wife, who has returned.*

[00:55:45–01:00:57] An often-seen element in any rom-com storyline involves the use of physical comedy, e.g. Meg Ryan famously faking an orgasm in a diner in *When Harry Met Sally* (1989), or Hugh Grant and Colin Firth fighting in *Bridget Jones's Diary* (2001), which was comically repeated in *Bridget Jones: The Edge of Reason* (2004). In *Finding Fatimah*, the scene at the jet-wash where Shahid unexpectedly meets his ex-wife, Fiza, successfully encapsulates this concept.

Shahid arrives at the petrol station with his sister Hafsah, with the intention of washing his van. Hafsah is left sitting in the van listening to the radio, whilst Shahid obtains a token to use the jet-wash. This is clearly diegetic music, and the radio station she is listening to in the van is also simultaneously playing in the garage shop where Shahid is. As the scene intercuts between the van and the shop, the perspective and treatment of the song changes accordingly (primarily, using the Spatial sound area to highlight the difference between the two locations).

In the van, the original song comes to a natural end. Hafsa leans forward as if to change the station on the radio and an upbeat song begins. The presence of the song does not suggest that the song is coming from the van's speakers; and as the pictures cut between the inside and outside of the van, the perspective of the music does not change. What then follows is a short 'feel good' musical montage of car-wash shots, with audible sound effects of the washing process: Hafsa pulls funny faces behind the glass as Shahid directs the hose at her; Shahid draws a funny face in the soap suds on the window.

We see and hear a car pull up behind Shahid's van at the car wash and this non-diegetic music continues until, after a car door slam and an impatient exclamation from its driver, the music abruptly stops with a vinyl record needle scratch effect. The line 'Can't you see I'm waiting?' from the female driver coming into view is enough for Shahid to recognize, without even seeing her, that this is Fiza, his fearsome ex-wife.

The musical motif that accompanies Shahid's exclamation of her name is that of a spaghetti western showdown (used in the Narrative sound area – the music is used here to pre-empt the violence that is about to follow); and the interchange between them becomes an escalating exchange of bruising, verbal blows.

Hafsa is puzzled by what she hears and gets out of the van to see who it is that Shahid is talking to; and a slanging match quickly descends into a physical struggle between Fiza and Hafsa. As Shahid watches the two women wrestling, the turmoil in Shahid's head is represented by the tocsin of a siren and a chaotic heavy metal riff (music is again used for its Narrative qualities).

When Hafsa learns that this woman is Fiza – the source of enduring misery to her elder brother and endless stress to their mother – she picks up the jet-wash lance and thoroughly douses the ex-wife, before throwing down the hose in disgust. The resulting image of a soaked and dishevelled Fiza, and the prospect of us seeing some kind of a comedic custard pie-style revenge from Fiza as she picks up the lance, increases this sense of slapstick. The pratfall, however, is that the 'out of time' bell on the jet-wash (Narrative sound area) rings just as Fiza pulls the trigger; and no water comes out of the hose; leading to Hafsa's infectious, mocking laughter.

However, the hilarity of the situation is quickly dispelled and counterpointed by Fiza, as she reveals some vengeful 'truths': her infidelity whilst she was married to Shahid, and her contempt of him mourning so deeply at the loss of his late father. The lightness of the comedy has been quickly, and cleverly, counterpointed by the heaviness of tragedy.

This is an emotionally complex scene where the core intended emotions are several-fold: initially, the scene is happy and upbeat, aligning with the Yellow (Sunny) quadrant for emotions such as amusement, joy and pleasure. Fiza's arrival and her verbal challenges to Shahid undoubtedly arouse in him emotions in the Blue (Alert) quadrant: feelings of regret, shame and disappointment. Finally, the scene moves towards the Red (Fiery) quadrant, with clear examples of disgust, contempt and anger (See Figures 7.2 and 7.3.)

Appendix 5

9.5 Typical feature film Sound Department hierarchy

Table 9.1 Typical feature film Sound Department hierarchy

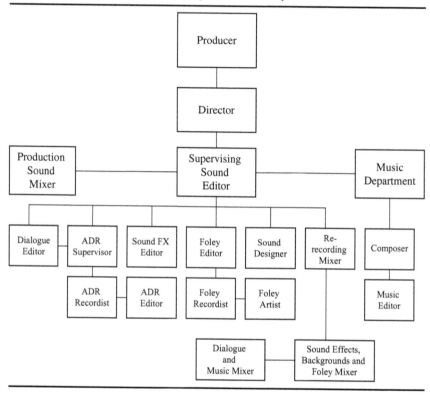

Appendix 6

9.6 Typical feature film post-production schedule – see Table 9.2

26 Week Post Schedule Template
13 weeks to Picture Lock
5 weeks Tracklay
4 weeks Mix
4 weeks Delivery
(Courtesy of Post-production Supervisor, Polly Duval)[7]

Appendix 7

9.7.1 Automatic Dialogue Replacement – ADR

The human voice is clearly special in film-making terms because it serves us as our primary form of communication. Therefore, it is an important soundtrack

Table 9.2 Typical feature film post-production schedule

Date	Post week	Activity
	1	**Editors assembly**
	2	**Directors Cut 1**
	3	**Directors Cut 2**
	4	**Directors Cut 3**
	5	**Directors Cut 4**
	6	**Directors Cut 5**
	7	**Directors Cut 6**
	8	**Directors Cut 7**
	9	**Directors Cut 8 – end of week: Screen to financiers**
	10	**Fine Cut**
		Lock for Preview
	11	**SOUND DEPT. STARTS**
		Prep for Preview – Temp ADR, Temp VFX, Temp Mix
	12	**PREVIEW**
	13	**Fine Cut**
		Final Financier Screening
		PICTURE LOCK END OF WEEK
	14	**Sound: Tracklay 1**
		Data extraction for DI conform
		DI conform begins
	15	**Sound: Tracklay 2**
		DI conform continues
	16	**Sound: Tracklay 3**
		Record ADR
		DI conform continues
		Start VFX delivery
	17	**Sound: Tracklay 4**
		Record ADR
		VFX delivery
		DI conform complete end of week
	18	**Sound: Tracklay 5**
		DI grade week 1
		End roller credits out for approval
	19	**Sound: Dialogue Pre-mix**
		DI grade week 2
	20	**Sound: Music Record & Mix**
		Sound: FX Pre-mix
	21	**Sound: Final Mix**
		HD mastering
	22	**Sound: Final Mix**
		Sound: Final Mix sign off
		Cutting room closes
	23	**Sound: M&E**
		Printmaster
	24	**HD audio laybacks & QC**
	25	**DCP clones and HD deliverables**
	26	**DELIVERY OF ALL FILM & VIDEO MATERIALS AND DOCUMENTATION**

element that needs to be carefully considered in any on-screen scene, especially when people are interacting together.

Even if there is no dialogue present, it can have the effect of leaving an audience anticipating what *might* be about to be said. This absence of speech might be being used for effect, to build the interest of the scene or to establish a sense of suspense, or even instead, to allow a sonic space for the background sounds to 'speak' alongside the pictures, conveying their own message. However, for the majority of narrative filmmaking it could be argued that the dialogue presents as the core sound-element that everything else is wrapped around.

It might seem an obvious thing to say, but simply hearing a character speak establishes an impression of them in the audience's mind: through their accent (it may be foreign or regional), through social or cultural markers (are they an articulate speaker, suggesting a well-educated or affluent background, or are they talking the talk of the 'street-wise', the less well-off or the less-privileged?). From the tempo of their speech we can determine if they are calm or agitated, and from their tone we can tell if they are nervous, confident, happy or angry. We can also make assumptions about their health and lifestyle: for instance, if they have a 'gravelly' timbre, this is synonymous with a smoker or a heavy drinker's voice. Finally, we can also tell whether the voice is old or young, high-pitched, masculine/effeminate and so on; all the things that an actor and director will work on when establishing a character, at the read-throughs and pre-production stages. Therefore, it is fair to say that a lot of information about a character is available to share with the audience, should it suit the story, even before we actually see them.

9.7.2 *Recording ADR*

The challenge facing the Dialogue Editor (and therefore the ADR Mixer) usually starts with determining whether the ADR loops are required to match directly alongside dialogue that was recorded on location; for instance it may only be a word or a sentence within a speech that is troublesome, and if that is the case – as opposed to replacing *everything* in the script – that means choosing microphones that can match the fidelity of the original location recording.

What this does *not* necessarily mean is simply using the same microphone that was used by the location recordist when capturing the production sound. For instance, if the dialogue of the 'good takes' was recorded on a boom microphone, that would almost certainly mean that they used an interference tube, hyper-cardioid or super-cardioid microphone. Unfortunately, unless you are recording in a large room with excellent acoustics, you may well find that this kind of microphone does not match as well as you hoped; especially if you are recording in a small voice booth. This is due to the physical characteristics that make it such a useful microphone on location. Put simply, these 'shotgun' microphones are designed to be used in wide open spaces, not somewhere small and enclosed. Put this kind of microphone too close to an actor indoors, and amongst other issues such as reflections and colouration from the rear lobe of the microphone, there is often an exaggerated bass frequency lift that is audible in the voice.

With the advent of reliable wireless microphone systems, small digital body-worn recorders, and the affordability of multi-track location sound recorders, actors are also routinely fitted with their own concealed lavalier microphone; a safety measure that means the Production Mixer's balance of the dialogue recording is often simply acting as a cutting guide, as all the microphones are recorded onto their own dedicated track of the recorder. In these circumstances, the same type of lavalier microphone should be used in ADR as on the original recording. But with the luxury of not needing to conceal the microphone, unwanted rustling noise from clothing can be avoided – a common reason for recording ADR in the first place. (The caveat to this is that successfully concealing a microphone, avoiding clothing rustle *and* capturing a usable, full-frequency voice recording on location is nothing less than an art in itself; literally invisible and widely unconsidered and under-valued by non-sound professionals.)

Almost always, my starting point for recording ADR in the studio is to record a matching type of lavalier microphone capsule on top of the actor's clothes, in the mid-chest region; and to also use a switchable-polarity, large diaphragm microphone set to an omni-directional polar pattern and placed about eight inches away from the *side* (not in front) of the actors face, to mimic the characteristics of the original boom track.

The most noticeable differences between the placement of the two microphones is largely due to the boom microphone being above an actor's head, whilst the lavalier microphone is below – resulting in very different tonal qualities. However, this new 'boom' track, utilizing the omni-directional microphone, is usually sufficiently neutral in tone to allow for its reasonably easy equalization by the Re-recording Mixer at the final mixing stage, which will ensure that the ADR is a good match to the original recording.

But until the ADR session actually begins and you hear your choice of microphones in action, there is always some uncertainty as to which microphone will work best; it absolutely depends on the characteristics of the actor's voice. Sadly, there are no hard and fast rules that work in each and every case.

9.7.3 The advantages and limitations of ADR

One of the biggest advantages of ADR is that you get time to concentrate purely on recording the sound without being conscious of affecting the functioning of other departments: on set, the Production Sound Mixer is under tremendous pressure not to hold up a shooting schedule; and whilst most sensible Directors understand that it is incredibly important to get the best original dialogue recordings that you can, and know that that can sometimes require time or a re-take to achieve, more often than not when the Recordist asks for another take, it's never as well received as, say, a focus puller asking for another take because a shot went soft at some point.

It is also worth mentioning that it is rare that any location Sound Recordist will be made to feel that they are granted the same amount of respect, support and understanding to do their job properly, as for instance a Director of Photography

(DoP), a Camera Operator, a Focus Puller, a Grip or even a Lighting Gaffer. And often, those location Sound Recordists considered to be the best by other members of a film crew are the ones that remain forever invisible and silent, quietly doing their job without disturbing anyone. Whilst possession of those Ninja-type skills are qualities that *all* Production Mixers aspire to, it is only an acceptable way to operate for as long as the Recordist delivers good sound to post-production. Their amiability on location is meaningless if a good deal of restoration or ADR work is needed in audio post-production, simply because they were too 'nice' to speak up when things were less than acceptable for the sound department on set. (Having been in that position many times myself, I know that it is a tough, but necessary, call to make. Standing your ground as a Production Sound Mixer, which actually means you are working for the benefit of the finished film, can be a very lonely and dispiriting place at times.)

One obvious disadvantage to ADR is that an actor will have been in a particular mindset when they originally performed their lines on camera, and possibly working with and alongside other actors whose performances inspired their own. ADR is generally an individual experience for actors; and that is why it is important to have them comfortable in their surroundings if they are to deliver a comparable performance to the one they gave weeks, or even months, before. The actor may also lack confidence in being able to 'post-sync' their scene (some actors love it, some loathe it), feeling sure that their performance on-set was not only as good as it could ever be, but also unsure if they can get back into the right frame of mind to deliver the emotion required. That is when the ADR Mixer and the Dialogue Editor's sensitivity, empathy, craft and people skills have to come to the fore; either with, or in the absence of, the film's Director.

9.7.4 Unexpectedly part of an ADR Masterclass

On one occasion, it was my great pleasure to be contacted to record a 'remote' ADR session[8] with an actor, by one of Skywalker Sound's Supervising Sound Editors, Jonathan Null[9] (more on this conversation at section 9.7.7). Whilst the actor to be recorded would come into my studio, Jonathan would listen-in and offer direction from the US and make a guide recording via the audio ISDN link between us at his end. The clean session audio recorded at my studio would be sent later to Skywalker via secure ftp, after it was neatly arranged in place on the project timeline and packaged as a .omf file.

The session itself required a vocal delivery from the actor that went from being an intimate, barely audible whisper, to a completely hysterical reaction to the devastating news of the loss of a parent. Not surprisingly, at first the actor felt that the emotions were simply too intense to be brought back up, out of context and in a strange studio, and was sure that it was not possible to replicate their performance. In fact, they refused to try.

Jonathan's tact and diplomacy in the way that he spoke with the actor ensured that not only was it possible, but by gently and respectfully guiding and coaching them through their role in the film, line by line, when it was felt by the actor that absolutely only one take of the 'hysterical' loop would be possible, we actually managed to record several versions.

The performances on all of the takes were superb and required enormous effort on behalf of the actor; and all were slightly different due to Jonathan managing to bring out different, subtle nuances in each new run of the loop, giving himself a good selection of takes to experiment with later, in the dialogue edit and pre-mix.

For my part, whilst I was undoubtedly awed listening to a master at work directing the actor, I was desperately conscious of the mechanics of me recording such a massive dynamic range, without a rehearsal to check the recording level. Instead, we just had to go for it, with me painfully aware that the last thing Jonathan needed was a distorted recording, after achieving what had seemed impossible at the outset – to even begin the recording session with this actor.

That example highlights the difficulty in trying to recapture a specific scene's mood and tone in ADR, and why I am a great fan of recording on location what I describe as 'Wild ADR'. Whenever I'm the Production Mixer recording on location, I try to arrange with the Director and First Assistant Director at the pre-Production stage, to be allowed time on-set to record any difficult scenes for sound once more, without the camera rolling, after the Director feels that they have got their best take for pictures and for performance.

This means allowing the boom operator one more take of the scene to place the microphone in the optimum position for sound, without worrying about staying out of shot. The actors are often so close in timing, rhythm and inflection to the 'circle take' shot on camera, that it can alleviate the need for ADR entirely; and if nothing else, it is reassuring to go into a dialogue edit knowing that you have this clean dialogue, in the same acoustic environment, as a fall-back. It takes only a short time on location to do this, but it can save many hours and significant extra costs in post-production.

9.7.5 ADR techniques

The ADR mixer has equalization more easily to hand on their mixing desk than the location recordist does, who should generally try to deliver a recording as 'flat' as possible to the dialogue editor, i.e. without using any equalization that will alter the original recording, other than an 80Hz LF roll-off to eliminate unwanted handling noise on the boom mic, for instance, or body-movement on the personal mic.

The untreated, original dialogue helps to give a consistent starting point for the ADR Mixer to match the ADR to; and this is an important consideration as Production Mixers become presented with an increasing number of tempting 'solutions' to extraneous location noise, through software and hardware additions to their recorders.[10] Also, in the more controlled environment of an ADR studio we can place a lavalier microphone on top of clothes, or position a microphone on a stand close to the actor, rather than relying on a buried personal microphone, or utilizing a shotgun microphone fully shielded against wind noise, being manoeuvred at the end of a long boom pole.

However, more than the absolute *fidelity* of the microphone, a listening-viewer's ear is subconsciously, yet highly, sensitive to the *change* in tone when switching between microphone types throughout a scene; and it requires great care and skill on behalf of the Re-recording Mixer to carefully merge the different microphone types across shot changes or dialogue crosses.

And it bears emphasizing: the use of bad ADR is as irritating to the listening-viewer as allowing poor original sound a place in the finished soundtrack: it severely diminishes the viewing experience and it is generally an unacceptable solution to an often avoidable situation.

The place where ADR can particularly enhance an actor's delivery is when that delivery is very low-level (or as extensively reported from the occurrence of well-publicised viewer's complaints, when it is *mumbled*).[11]

On set, the Production Mixer is constantly battling against the signal-to-noise ratio – but in the controlled conditions of a sound studio, this is much less of an issue. To aid intelligibility, in audio post-production we can provide help to some voices by giving careful and selective boosting in the 2 kHz to 6 kHz range, an area that the ear is particularly sensitive to. Too much in this range however will lead to a thin 'telephonic' effect. Some Low–Mid boosting around 250Hz can also provide a sense of 'weight' or depth to a voice. But these are only starting points. Every voice, clichéd as it might sound, is as different as a finger print; and it is perfectly possible to find yourself surprised at how good someone's voice sounds on a small-capsule, lavalier microphone, compared to the sound from a shotgun or a large diaphragm microphone.

Successful ADR recording requires excellent organizational skills and manual dexterity on the part of the ADR Mixer: each take must be clearly identifiable and each microphone type should be clearly labelled to relate to the clip names being used, to enable the dialogue editor to easily find chosen takes later.

It is also useful to have to hand background atmosphere tracks that can 'sit' the ADR lines into a scene during playback, so the Director and Dialogue Editor can quickly hear the lines both clean and with a more realistic background played around them.

An ADR session can be a very busy time for the ADR Mixer, who needs to make it all look very casual and relaxed whilst ensuring that – just like on-set sound recording – neither the actor nor the Director is ever kept waiting by the extensive logistics of operating the ADR process.

9.7.6 *The privilege of watching a performance*

One of the enduring thrills of me recording ADR is witnessing first-hand how an actor can bring a character to life. When I am engaged to record ADR on a film that I am not dialogue editing, or indeed have no other input to, the pictures received from the cutting room are of random, unrelated scenes and usually sent with two totally separate audio tracks: clean dialogue and clean effects; and at this stage, in its component form, this disparate content is disjointed and mechanically separated from the whole that it will eventually become after final mixing.

It is also an unknown quantity in terms of approach and performance, and on a 'remote' session, having the Director come on the line as a disembodied voice to direct the artist, only adds to the artificiality of the situation, and highlights the potential loneliness of the long-distance actor. Nowadays, video calls help.

It is perhaps unfair to single out any one artist as being more impressive than any other, because frankly, 'in the booth', they are all remarkable to me; but there is one particular memory that lingers of a beautiful, haunting performance, delivered many months after its filming.

Julie Christie starred opposite John Hurt and Shia LaBeouf (the two men playing old and young versions of the same character) in the *Love of Violets* scene within the single unified 'collective feature film', *New York, I Love You* (2008), an ensemble movie brought together by Producer Emmanuel Benbihy (Zuckerman, 2009).

Miss Christie played a former opera singer returning to her favourite Manhattan hotel, reminiscing with the young bell-boy, played by Shia; and it had been shot knowing that all of her lines would require replacing. The delicate, tender script was written by Anthony Minghella, who was also ready to Direct; however just before shooting began he was taken ill, and so he passed that role to Shekhar Kapur; and so it was that it was he who conducted the ADR session with us in Birmingham, from a Los Angeles studio.

Listening and watching Julie through the glass, as she moved and gestured through her, at times, quiet and intimate delivery, was to eavesdrop and witness an incredible artist at work; testament to the fact that Oscars are not generally awarded to actors who are not masters of their craft.[12]

Well received for her performance, Julie beautifully revealed every aspect of Minghella's poetic script. In reviewing her role in the film, Emanuel Levy wrote:

> through the veil of a curtained window, events play out that might be real, imaginary or a heartrending tear in the fabric of time. (Levy, 2009)

Tragically, Anthony Minghella died between him writing the script and filming being completed; which added an acute poignancy to the work we were doing. Unbeknown to him at the time, film critic David Zuckerman later captured what we had all felt that day:

> the aura of death surrounding both the narrative and hanging over the production raises this vignette to a higher level. (Zuckerman, 2009)

It was a privilege for me to be a part of the proceedings.

9.7.7 *A lesson on humbleness, collaboration and motivation*

During my initial conversation with Skywalker's Jonathan Null, we spoke about the scenes to be recorded and I asked Jon what microphones he wanted to use, to which he replied, 'what do *you* think we should use?'

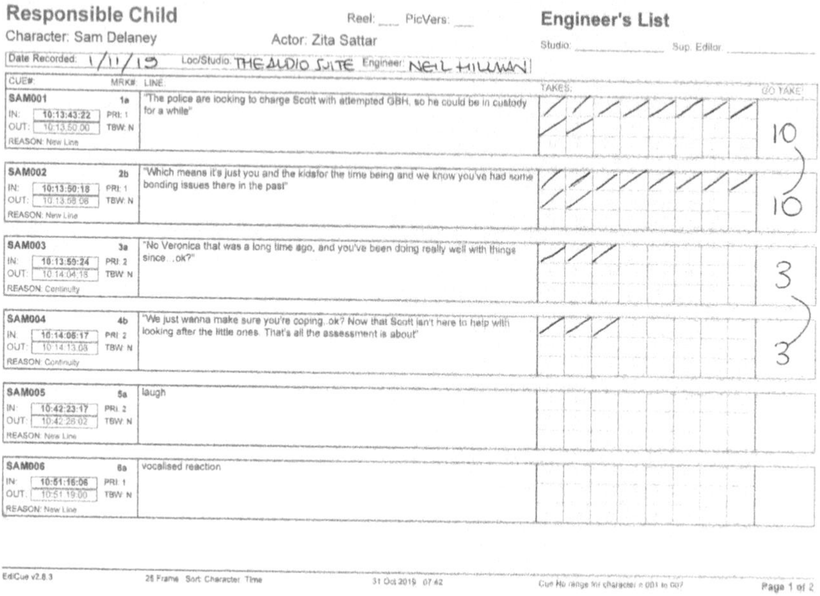

Responsible Child Reel: ___ PicVers: ___ **Engineer's List**

Character: Sam Delaney Actor: Zita Sattar Studio: ________ Sup. Editor: ________

Date Recorded: 1/11/19 Loc/Studio: THE AUDIO SUITE Engineer: NEIL HILLMAN

CUE#	MRK#	LINE	TAKES	GO TAKE
SAM001 IN: 10:13:43:22 PRI: 1 OUT: 10:13:50:00 TBW: N REASON: New Line	1a	"The police are looking to charge Scott with attempted GBH, so he could be in custody for a while"		10
SAM002 IN: 10:13:50:18 PRI: 1 OUT: 10:13:58:06 TBW: N REASON: New Line	2b	"Which means it's just you and the kidsfor the time being and we know you've had some bonding issues there in the past"		10
SAM003 IN: 10:13:59:24 PRI: 2 OUT: 10:14:04:16 TBW: N REASON: Continuity	3a	"No Veronica that was a long time ago, and you've been doing really well with things since...ok?"		3
SAM004 IN: 10:14:06:17 PRI: 2 OUT: 10:14:13:08 TBW: N REASON: Continuity	4b	"We just wanna make sure you're coping..ok? Now that Scott isn't here to help with looking after the little ones. That's all the assessment is about"		3
SAM005 IN: 10:42:23:17 PRI: 2 OUT: 10:42:26:02 TBW: N REASON: New Line	5a	laugh		
SAM006 IN: 10:51:16:06 PRI: 1 OUT: 10:51:19:00 TBW: N REASON: New Line	6a	vocalised reaction		

EdiCue v2.8.3 25 Frame Sort: Character Time 31 Oct 2019 07:42 Cue No range for character = 001 to 007 Page 1 of 2

Figure 9.2 Example of the Engineers List for an ADR cue sheet

I suggested that a shotgun mic and a lavalier would almost certainly have been used on-set, so we would obviously need to use both of those types, but how did he feel about me also offering an omni-directional, large diaphragm microphone at the side of the artist's head? Jon said he was not familiar with that kind of technique, but he really liked the idea of it; so, could we please add that to the recording tracks we would make, on the day of the session?

What I learned from that exchange, and by the way we worked together on the session itself, was that even though Jon was a Dialogue Editor at the very top of his game, working for one of the world's leading film Directors at one of the world's finest sound studios (and with multiple awards and many credits for films I'd enjoyed, such as *Bicentennial Man* (1999), *The Polar Express* (2004) and *How to Train Your Dragon* (2010)), he still took the time to ask for my thoughts, he listened attentively and he made me feel that my contribution was valued. Quite simply, he was inspirational to work with and I remain grateful for the opportunity to be part of his sound team on that particular film.

Notes

1 Whilst what constitutes a low budget film is dependent on its country of origin and genre, in the UK a production is generally considered to be so if it has a budget of less than £1 million. *Finding Fatimah* was made for £325,000.

2 One emotional signpost for Fatimah's anger was a signature heavy metal musical riff that Re-recording Mixer Pip Norton carefully emphasized in a different way each time it occurred, to prevent it becoming too obvious and repetitive.

3 Anna Sulley: www.imdb.com/name/nm1274490/.

4 Sue Malpass: www.imdb.com/name/nm1583403/.

5 The Powell theatre at Pinewood studios is one of the UK's largest mixing theatres. Developed in 1935 as a film studio and backlot on the site of Heatherden Hall by millionaire builder Charles Boot and wealthy flour industrialist J. Arthur Rank, Pinewood was built on the lines of the Californian film studios of the day; and its name was meant to be deliberately reminiscent of the word 'Hollywood'. It has arguably been the most prolific of the British film studios and it is synonymous with the production of the Carry On comedy series and James Bond franchises.

6 Pip Norton: www.imdb.com/name/nm0636276/?ref_=fn_al_nm_1.

7 Polly Duval: www.imdb.com/name/nm0245082/.

8 Remote ADR is when the artist and the recording take place in one studio, whilst simultaneously being recorded in another studio (often in another country), with the Dialogue Editor and Director listening in. It is also usual to synchronize pictures via timecode over the remote audio link (previously using ISDN, now increasingly an Audio over IP (AoIP) application) to gauge ADR delivery against the original performance. It is a complex process that must appear seamless to both the actor and the Director, with no technical difficulties impacting on the session's smooth running. It requires dexterity and vigilance by the ADR Mixer at both ends to ensure that the session flows in a transparent and timely manner.

9 Jonathan Null: www.imdb.com/name/nm0637871/?ref_=fn_al_nm_1.

10 Extraneous, background noise from air-conditioning, lighting hum etc. are the bane of a Production Mixer's working life. However, the solutions manufacturers offer with portable versions of audio post-production tools, e.g. the Cedar Audio DNS-2, or built-in recorder facilities such as the Sound Devices *NoiseAssist* application, require judicious use (and a clean parallel recording) if they are not to fatally prejudice the level of restoration that is possible in a controlled, audio post-production environment.

11 The *Daily Telegraph* reported that the BBC adaption of *A Christmas Carol* (2019) was the latest in a string of television dramas plagued by unfeasibly low-level dialogue delivered by some of its actors; with questions asked about the decisions made regarding the dialogue at the post-production and final mixing stage. A sound expert commented: 'For the avoidance of doubt, mumbled dialogue is not the audio equivalent of dark, moody pictures' (Carpani, 2019). It was subsequently re-mixed.

12 Julie Christie has won one Oscar, for her role in *Darling* (1965), and has received a further three nominations for *McCabe & Mrs. Miller* (1971), *Afterglow* (1997) and *Away from Her* (2006).

10 Credits

Sound for Moving Pictures

by

Dr. Neil Hillman

With special thanks to

Dr. Sandra Pauletto

and

Dr. Michael Filimowicz

Commissioning Editor – Hannah Rowe

Editorial Assistants – Shannon Neill and Adam Woods

Production Editor - Cathy Hurren

Editors – M. Raymond Izarali and Dalbir Ahlawat

Typesetting – Deanta Global Publishing Services, Chennai, India

Project Manager – Keith Emmanual Arnold

Copyeditor – Florence Production Ltd

Indexer – Lisa Stumpf

Design – Matthew Willis

Data Graphs – Samuel Hillman

Technical Figures – Dominic Osborne

Marketing – Edward Hall

Publicity – Edward Hall

Sales – Peter Williams

Glossary

AAF   Advanced Authoring Format (file extension .aaf). Successor to the OMF2 file as the de-facto audio media transfer tool between picture editing systems and digital audio workstations, it allows for greater file size and increased metadata than OMF2.

Abstract Sound   One of Hillman's Four Sound Areas concerned with sounds that have a less codified and clear meaning, such as atmospheres, backgrounds, room tones, sound effects and music.

Actuality Sound   Sound associated with pictures at the time of filming, e.g. the original location sound.

Adam Smith 2600   An external analogue timecode generator/reader used to synchronize video and audio devices. Popular in the 1980s, it was superseded by digital technology.

ADR   Automated Dialogue Replacement; recording replacement speech from actors in time with the original pictures.

Akashic records   From a theosophical belief system, they are said to be a compendium within which all human events, thoughts, words, emotions and intent ever to have occurred in the past, present or future are said to be contained.

Alert   The third quadrant of emotions in Hillman's Mix Disc. Coloured blue.

Ampex   An American electronics company founded in 1944 by Alexander M. Poniatoff, the company produced significant, cutting-edge audio and video recorders between the 1940s and 1990s. The word AMPEX is a play on the founder's name: Alexander M. Poniatoff Excellence.

Atmospheres   Sounds that work to define the time and place of a location.

Audio over IP (AoIP)   A standard procedure for the distribution of digital audio across an Internet Protocol (IP) network, such as the internet, or a similarly configured computer network.

Auteur   Derived from the French 'author', it is a term applied to Directors who provide the driving force behind the artistic endeavour and production processes of a feature film, in a similar way to that of an author producing a written work of fiction. Notable names include Alfred Hitchcock, Robert Altman and Federico Fellini, amongst many others.

Backgrounds   Sound effects used in a feature film to help present the time and place of a location; also referred to as atmospheres or 'atmos'.

BBC   British Broadcasting Corporation.

Bed   Underlying composite sounds, often atmospheric in nature.

Boom operator   A highly responsible position in the location (production) sound team that requires the accurate positioning of a microphone at the end of a telescopic pole (i.e. the 'boom') close to actors' mouths, without the microphone being seen within the camera frame. Also fits actor's lavalier-style personal radio-microphones. Can be credited as 1st Assistant Sound. Often known as a 'boom swinger', on set.

Bus   A prescribed or assignable path to aid the routing of signals within audio equipment such as mixing desks.

Channel 4   UK broadcaster funded by commercials.

Channel 5   UK broadcaster funded by commercials.

Circle take   Industry term used to denote the best take of a filmed scene or the preferred recording of an ADR loop.

Close up (CU)   A picture department term for when a subject's face is tightly framed. Variations include big close up (BCU) and medium close up (MCU). One of the three standard camera framings of Close up (CU), Mid-shot (MS) and Long shot (LS).

Colourist   The person who digitally processes the final image of a project.

Compressed   An audio signal or file in which the amount of data in the recorded waveform is reduced by variable amounts, usually with some loss of quality.

Compressor   A hardware or software device used to lessen the dynamic range between the loudest and quietest parts of an audio signal.

Conform   Re-assembly of the sound elements to match a new version of the picture edit; or the assembly of sound elements from their original sources to match their chosen location in a picture edit.

Convolution reverb   A process that digitally simulates the reverberation characteristics of a physical or virtual space through the use of multiple and various software profiles.

Cut   Predominantly referring to pictures, it can be a single edit in a scene or refer more generally to a version of picture edit, e.g. the 'rough cut' versus the 'fine cut' or 'locked cut'.

DAW   Digital Audio Workstation; can be solely software-based, or a combination of software with an external hardware controller.

Deliverables   The specified master materials required by broadcasters and distributors; this usually includes several 'sub-masters' such as an M&E track (see *M&E*) as well as the master soundtrack for a TV programme or feature film.

DI   Shortened name for Digital Intermediate, the finishing process where a film's colour and image characteristics are manipulated in preparation for final delivery.

Diegetic sound   Sound that originates from a source that can be seen or is directly connected to the film story. It covers both on- and off-screen sounds.

Digi-neg   Short for 'digital negative'. An inkjet-printed copy of a film negative, usually larger than the original source. The file used to create the digi-neg may come from scanned film or from a digital camera image.

Dubbing Mixer   British film and television term for the person responsible for ensuring that the finished film or television soundtrack is correct both technically and stylistically. Also known as a Re-recording Mixer.

Dubbing Mixer Units (DMU)   Fictional sub-divisions of time that flexibly encapsulate the time taken to achieve a task in the mixing studio.

EQ   Equalization; the process of adjusting the balance between frequency components within an electronic signal.

Extra   More correctly known as a 'supporting artist' they perform non-speaking roles that make scenes in films and TV programmes look authentic, such as diners in a restaurant or passers-by on the street. A supporting artist may be given a line, but they then become a member of the cast, with a 'role title' such as waitress or receptionist.

Fiery   The fourth quadrant of emotions in Hillman's Mix Disc. Coloured red.

Film-out   A broad picture post-production term that encompasses the conversion of frame rates, colour correction, as well as the creation or conversion of regional and presentation standards, e.g. PAL, NTSC, HD.

First Assistant Director   Also known as the 1st AD. They are responsible to the Director for the smooth running and facilitation of the on-set filming. It is important to understand that they are judged by their ability to keep the crew and a filming schedule on time, however ambitious, as opposed to contributing to the quality of the final product. Also, unofficially responsible for setting the tone for the way in which the sound department are respected on location.

Five   UK broadcaster funded by commercials; re-brand of original 'Channel 5'.

Foldback   A feed of an audio mix sent to a performer, as opposed to the audience.

Foley   Sound effects such as footsteps and clothing rustle recorded in synchronization with edited pictures in post-production. Named after Jack Foley, who was the head of the sound effects department at Universal Studios.

Footage counter   Before timecode was introduced to film production, the units used for timeline positioning in post-production were the imperial measurement of feet, where 1 foot of 35mm film contained 16 frames (12 inches = 1 foot). Feature film cameras traditionally shot at 24fps.

Fps   Frames per second; interval speed of recording recorded pictures.

Fs   Sampling frequency for digital audio; typically, 44.1, 48 or 96kHz

Ftp   Abbreviation of File Transfer Protocol. A computer language for transferring data files between computers.

Grams Operator   On live television studio shows, the person who operates the play-in machines for music and sound effects.

HD   High Definition.

HDTV   High Definition Television.

Helical scan   Invented by the Ampex Corporation in 1956, it is a method by which high-frequency signals are recorded onto magnetic tape by machines such as video and audio recorders. The mechanically complex helical scan technology enabled the first recordings of video pictures in real time.

HPF   High Pass Filter.

IMDb   Shortened version of the name Internet Movie Database; an online reference website for looking up cast, crew and production data on feature films and television programmes. Extensive but not definitive.

Indie   Short for Independent. An independent film is produced outside the major film studio system, being produced and distributed by independent entertainment companies.

ISDN   Integrated Services Digital Network; a standard for the transmission of voice, video and data across the public switched telephone system. Through the use of dedicated codecs, a means of high quality audio link between studios, or from Outside Broadcast locations.

ITV   Independent Television; UK broadcast network funded by commercials.

Lavalier microphone   A small capsule microphone that can be concealed under clothing or worn visibly, i.e. on a lapel or tie. May be cabled if the subject is seated in a studio discussion or attached to a radio transmitter pack that is worn if the subject is moving. Also known as a 'personal mic' or a 'radio mic'.

Layback   The transfer of an audio mix to a master (or sub-master) medium.

LFE   Low Frequency Effects.

Limiter   A compressor with a high ratio, typically greater than 10:1; allows signals below a given power to pass, whilst attenuating those above.

Locked cut   The final version of pictures that will receive no further picture editing, i.e. the pictures are now locked to the sound.

Long shot   Picture department term for a camera shot taken from a considerable distance from the action, where people can become indistinct shapes. Variations include extreme long shot and medium long shot. One of the three standard camera framings of Close up (CU), Mid-shot (MS) and Long shot (LS).

Loops   The lines that an actor replaces after filming is complete, by listening repeatedly to a single line of dialogue ('looping the audio') and then speaks in time with the on-screen character. See *ADR*.

LoRo   Left only/Right only refers to a stereo downmix from a 5.1 signal and is also known as an 'ITU Downmix'.

Loudness   The perceived level of an audio signal; governed in the UK and European broadcasting by the EBU R128 standard and in the US by ATSC A/85.

LPF   Low Pass Filter.

LtRt   Left total/Right total; the nomenclature for a stereo signal encoded to Dolby Pro-logic; an encoded matrix of four to six discrete source channels. When decoded with Dolby Pro-Logic, the playback channels are Left, Centre, Right, Surround (L, C, R, S); when decoded with Dolby Pro-Logic II, the playback channels are Left, Centre, Right, Left surround, Right surround, LFE (L, C, R, Ls, Rs, LFE).

LUFS   Loudness Units relative to Full Scale; a measurement standard for loudness in television broadcasting world wide. See *loudness*.

MADI   Multichannel Audio Digital Interface.

M&E   Music and Effects. A standard deliverable for foreign or re-versioned programmes, it contains the music and effects tracks only mixed together (i.e. the full mix minus the original dialogue) and enables the replacing of original dialogue with another language.

Media Pool   The repository for audio and picture media within a non-linear editing system or application.

Mid shot (MS)   Picture department term for a camera shot used to show both actors' facial expressions and body language, whilst also allowing a view of the background. Subjects and background equally share the frame. Variations include a single MS, a group MS, a two-shot MS, and an over-the-shoulder MS. One of the three standard camera framings of Close up (CU), Mid-shot (MS) and Long shot (LS).

Mild   The second quadrant of emotions in Hillman's Mix Disc. Coloured green.

Mix Disc   A colour-coded circle subdivided into four coloured quadrants, where each colour represents one of four general emotional states. Used in spotting sessions and when planning sound design, it forms a part of Hillman's Four Sound Areas framework.

Mix Wheel   An adaptation of the Geneva Emotion Wheel. Used in spotting sessions and when planning sound design, it forms a part of Hillman's Four Sound Areas framework.

MPSE   Motion Picture Sound Editors; US-based honorary organization.

MS (Middle and Side)   A stereo recording technique invented by pioneer engineer Alan Blumlein in the 1930s and still a popular form of stereo encoding in broadcast applications because of its inherent mono compatibility and adjustable stereo imaging. Achieved with two microphone elements (typically a cardioid capsule for the 'middle' signal and a figure-of-eight capsule for the 'side' signal).

Mute   Switch used to silence the output of a signal path.

Nagra   Standard feature film industry, high-quality quarter-inch magnetic tape recorder, used extensively as the portable machine for recording location dialogue and effects from the 1950s to the 1990s. Various models between versions I and IV were popular: Mono (I–III), Stereo with 50Hz pilot tone (IV–S) and Stereo with on-board timecode generator (IV–S TC).

Narrative Sound   One of Hillman's Four Sound Areas concerned with sound that carries direct communication and meaning. Examples include dialogue and commentary as well as symbolic and signalling sounds such as ringtones and sirens; it also includes music with a clearly defined meaning.

Negative film   The name for a type of high-quality photographic film stock that has its colours inverted after development. A second process called 'making a print' is used to obtain the final images. (See also *reversal film*.)

Ninja   From the Japanese word for spy, it is used in popular culture as a term for someone who acts with stealth and great skill to achieve good results.

Non-diegetic sound   Sound whose source is not seen on-screen, nor meant to be thought of as sound connected to the on-screen action. (The opposite to Diegetic sound.)

Octave   A doubling in frequency.

OMF   Short for Open Media Format (type OMF1 and later type OMF2, file extension .omf). It is a file type for audio media exchange between picture editing systems and digital audio workstations. Long superseded by its successor the AAF (file extension .aaf) it steadfastly remains in use as a media transport device. See *AAF*.

PA   Public Address; the loudspeaker feed to the audience.

Peak   A signal's maximum output level.

Peaking   A signal reaching maximum output level.

PFL   Pre-Fade Listen; a monitoring facility allowing a signal to be monitored without fading it up in a mix.

Pilot tone   A means of synchronization on early Nagra quarter-inch magnetic tape audio recorders (pre-timecode generator recording). A 50Hz signal would be recorded between the two audio tracks on the portable machine, which would be used as a reference signal to regulate the playback speed on other suitably equipped playback devices.

Plug-in   A software accessory that sits within a DAW operating system but provides its own user-interface. Used to analyse, transform or generate new audio samples.

PPM   Peak Programme Meter; standard broadcast level measurement tool.

Pre-Mix   The mixing of all the edited sound elements for, e.g. the dialogue, music or sound effects into one single clip ahead of the final mix, so that the final mix is accomplished quickly and more easily, with a smaller track-count.

Printmaster   The final, composite sound mix of dialogue, music and sound effects that is transferred directly to the final print.

Production Mixer   The person responsible for recording the sound on location; Head of the location Sound Department, assisted by 1 or 2 boom operators (credited as 1st Assistant Sound) and a more junior assistant (credited as 2nd Assistant Sound).

Production sound   The sound recorded on location by the Production Mixer.

QC   Quality Control.

Reel   Although originally referring to 16mm or 35mm film production, modern non-linear post-production will often still break down a feature into segments of a roughly equivalent time to a 35mm reel (approximately 20 minutes). A standard-length feature film is known as a 'five-reeler' (as opposed to a 'short film' which is 20 minutes or less in duration).

Re-recording Mixer   The person responsible for ensuring that the finished film or television soundtrack is correct both technically and stylistically. An alternative term for the British title of Dubbing Mixer.

Reversal film   The name for a type of photographic film stock that produces a viewable (positive) image as soon as it is developed. Quicker and cheaper than Negative film stock. (See *negative film*.)

RMS   Shortened term for Root Mean Square. An audio RMS meter can be used as an approximation of the way the human ear perceives sound levels as ears do not typically perceive sharp peaks to be as loud as they actually are.

Rough cut   The point in the picture editing process where the film first starts to resemble its intended layout. The changes resulting from observing the rough-cut lead towards several further refining picture edits (known as 'fine cuts') before the final version of the pictures is arrived at (known as the 'locked cut').

RX   Shortened term for Rehearse, Record, Receive or Receiver.

Score   The music used within a film. Often wrongly described as 'the soundtrack', which is only used to describe the music used in a film when it is commercially released to the public, e.g. as a 'soundtrack album'.

Scratch mix   A fast mix of early work-in-progress material for investors or distributors to view. The audio equivalent of the picture 'rough cut'.

SD   Standard Definition.

Shotgun microphone   A uni-directional type of microphone extensively used to record dialogue on location, fixed to a boom pole. Usually of an interference-tube design for its focussed directivity and off-axis rejection qualities. The term 'shotgun' refers to the visible similarity to the firearm; a derivative term from the longer version 'rifle microphone', e.g. a shotgun Sennheiser 416, a rifle Sennheiser 816.

Sizzle   A sizzle (or 'sizzle reel') is a dynamic, fast paced highlights package of usually less than 5 minutes duration, designed expressly to excite and arouse interest in a product, a service, a brand or a person.

Solo   The ability to isolate a signal within an audio mix.

Sound Designer   The person responsible for the overall style of the sound of a film. In some cases, the Sound Designer will supervise both the sound editing and re-recording stages of audio post-production. Sometimes included in credits simply as the Supervising Sound Editor.

Sound Follower   Also known as separate magnetic, sepmag, magnetic film recorder or mag dubber, it is a machine that records and plays back sound from film stock with a magnetic audio track. The magnetic film stock is locked or synchronized with the film containing the picture.

Sound Supervisor   Most usually used in the television industry to denote the person mixing the sound and in overall charge of a programme's audio output.

Soundtrack   The full audio mix of a feature film.

Source music   Music heard by the on-screen characters. See *diegetic sound.*

Spatial Sound   One of Hillman's Four Sound Areas concerned with the positioning of sounds within a three-dimensional soundfield and the space placed around the presented sound.

Spotting   A 'session' where the Sound Designer or Supervising Sound Editor compiles a list of the sound effects or music required, scene by scene, throughout a film; also used as a time for reviewing a film with the Director to determine or explain what work will be required on the soundtrack.

Standard deviation   Denoted by the symbol σ, standard deviation is a number used to tell how far measurements in a group are spread out from the average value (the 'mean'). Low standard deviation means that most of the numbers are close to the average, whilst high standard deviation means that the numbers are more spread out.

Stems   The component parts of a mix, usually comprising separately of dialogue, music and sound effects that when combined make up the full final mix of a film or television programme. The industry shorthand term *Wide Stems* is used to denote the stems for the biggest deliverable format of the project, e.g. for a 7.1 project the wide stems would consist of *Left, Centre, Right, Left surround, Right surround, Left Rear surround, Right Rear surround, LFE*.

Steenbeck   Flatbed 16mm film editing table, e.g. Series-00 and -01 models.

Strowger   semi-automatic device for routing audio signals from one fixed destination to another. First used in telephone exchanges and adapted for use in broadcast television studios. Superseded by digital switches.

Studer 24 track   A professional studio 2-inch magnetic tape recorder offering 24 individual recording tracks.

Sub   Subwoofer; speaker dedicated to low frequency signals.

Sunny   The first quadrant of emotions in Hillman's Mix Disc. Coloured yellow.

Supervising Sound Editor   The person in charge of the sound editorial process, including dialogue, Foley, and sound effects editing. Can sometimes be used as an alternative term for Sound Designer.

Tableau Vivant   A theatrical term referring to actors posing silently to represent a scene, painting or sculpture. Literally, 'a living picture'.

Technical Review   The Quality Control process undertaken for every broadcast television programme.

Temp Mix   A quick mix made during the audio post-production process to help determine how things are working together in the soundtrack.

Temp Music   Music used as a placeholder in the soundtrack whilst the final choice of music is selected or written.

Temporal Sound   One of Hillman's Four Sound Areas concerned with the evolution in time of the sound design. Its characteristics are rhythm, pace and punctuation. This area can include music, sound effects and voice.

Timeline   The visual display of edited sequences on a non-linear editing system, arranged in chronological order.

Track laying   The term used to describe the creation, selection and positioning of audio by the Sound Editor/Sound Designer on a DAW, in advance of delivery to the mixing stage.

Two-shot   A framing of the camera to include two featured actors.

TX   Shortened term for Transmission, Transmit or Transmitter.

Uncompressed   The most accurate digital representation of a sound wave, as epitomised by the Wave, Broadcast Wave, AIFF and mxf file formats.

UHD   Ultra High Definition (in picture terms, sometimes incorrectly referred to as '4K') although often advertised as one-and-the-same, at 4,096 by 2,160 (an aspect ratio of 1.9:1) cinema 4K resolution is greater than that used for

Ultra High Definition (UHD) television at 3,840 by 2,160 (a smaller aspect ratio of 1.78:1).

VFX Shortened term for visual effects.

VT Originally Video Tape or Video tape machine; but still used as a term for the playing in of pre-recorded pictures (e.g. title sequences) from solid-state hardware or software devices in a live environment. Remains in common use by Directors calling 'Roll VT' on live television productions.

References

American Cinematographer (2018), Star Wars: The Last Jedi, *American Cinematographer Magazine*, Volume 99, Number 2, p. 45.

Associated Press (2019), UEFA Orders Montenegro to Play Game without Fans over Racism, at www.independent.com.mt/articles/2019-04-26/football/UEFA-orders-Montenegro-to-play-game-without-fans-over-racism-6736207235, accessed 05-05-20.

Attali, J. (1985), *Noise: The Political Economy of Music* (Trans. Brian Massumi), Manchester: Manchester University Press.

Bachorowski, J. (1999), Vocal Expression and Perception of Emotion, *Current Directions in Psychological Science*, 8(2), pp. 53–57.

Banse, R. & Scherer, K. (1996), Acoustic Profiles in Vocal Emotion Expression, *Journal of Personality and Social Psychology*, 70(3), March, pp. 614–636.

Barsam, R. & Monahan, D. (2009), *Looking at Movies: An Introduction to Film* (3rd Edition), New York: W. W. Norton & Company.

BBC (2014), Glasgow 2014: Commonwealth Closing Ceremony Watched by 6.8m, at www.bbc.co.uk/news/entertainment-arts-28640829, 04/08/14, accessed 07-08-14.

Blake, L. (2004), George Lucas – Technology and the Art of Filmmaking, *Mixonline* 11-01-2004, at www.mixonline.com/news/profiles/george-lucas/365460, accessed 04-06-17.

Bradshaw, P. (2014), *Here and Now* Review, *The Guardian* 03-07-14, at www.theguardian.com/film/2014/jul/03/here-and-now-review-teen-romance-lisle-turner-lauren-johns, accessed 22-10-14.

Bradshaw, P. (2017), *Finding Fatimah* Review – Faintly Desperate Dating Comedy, at www.theguardian.com/film/2017/apr/20/finding-fatimah-review-faintly-desperate-dating-comedy, accessed 19-05-20.

Bresson, R. (1977), *Notes on Cinematography* (Trans. Griffin, J.), p. 28, New York: Urizen Books.

Brown, A. R. & Sorenson, A. (2009), Integrating Creative Practice and Research in the Digital Media Arts, in *Practice-Led Research, Research-Led Practice*, Smith, H. and Dean, R. T. (eds), pp.153–165, Edinburgh: Edinburgh University Press.

Brown, M. (2013), Margaret Thatcher Was the Architect of Controversial Changes to TV and Press, at www.theguardian.com/media/media-blog/2013/apr/12/margaret-thatcher-television-press, accessed 15-03-20.

Candy, L. (2006), *Practice Based Research: A Guide*, University of Technology, Sydney Creativity & Cognition Studios Report November 2006, at www.creativityandcognition.com/resources/PBR%20Guide-1.1-2006.pdf, accessed 05-06-20.

Carpani, J. (2019), BBC Criticised for 'Mumbling' Adaptation of a Christmas Carol, *Daily Telegraph* 23-12-19, at www.telegraph.co.uk/news/2019/12/23/bbc-criticised-mumbling-adaptation-christmas-carol/, accessed 01-06-20.

Carr, J. (2015), *The Normalisation of Surveillance Through the Prism of Film: A Practice-Based Study*, PhD thesis, University of York.

Chion, M. (1994), *Audio-Vision: Sound on Screen*, New York: Columbia University Press.

Clarke, W. V. (1956), The Construction of an Industrial Selection Personality Test, *Journal of Psychology: Interdisciplinary and Applied*, 41(2), at www.tandfonline.com/doi/abs/10.1080/00223980.1956.9713011, accessed 01-05-20.

Clore, G. L. (1994), Why Emotions Are Never Unconscious, in *The Nature of Emotion*, Ekman, P. & Davidson, R. J. (eds), pp. 285–299, New York: Oxford University Press.

Cohen, A. J., MacMillan, K. A. & Drew, R. (2006), The Role of Music, Sound Effects & Speech on Absorption in a Film: The Congruence-Associationist Model of Media Cognition, *Canadian Acoustics*, 34, pp. 40–41.

Coleman, T. (2017), *Finding Fatimah* Review, *Total Film Magazine* April 2017, at www.gamesradar.com/movies-to-watch-21-april-2017/, accessed 19-05-20.

Collins, K. (2011), Making Gamers Cry: Mirror Neurons and Embodied Interaction with Game Sound, *Proceedings of the 6th Audio Mostly Conference: A Conference on Interaction with Sound*, pp. 39–46, New York: ACM.

Cowie, P. (1990), *Coppola*, New York: Da Capo Press.

Cowie, R. (2000), Describing the Emotional States Expressed in Speech, *ISCA Workshop on Speech & Emotion*, Northern Ireland, pp. 11–18.

Damasio, A. (2000), *The Feeling of What Happens*, London: Vintage.

Darwin, C. (1890), *The Expression of the Emotions in Man and Animals*, New York: D. Appleton & Co.

Deleuze, G. & Guattari, F. (1980), *A Thousand Plateaus*, London: Continuum.

Denison, C. (2017), Ultimate Surround Sound Guide: Different Formats Explained, at www.digitaltrends.com/home-theater/ultimate-surround-sound-guide-different-formats-explained/, accessed 16-05-17.

Deutsch, S. (2018), Quoting from private correspondence to the author.

Dixon, T. (2003), *From Passions to Emotions: The Creation of a Secular Psychological Category*, Cambridge: Cambridge University Press.

Dream Team, F. C. (2016), West Ham Accused of Playing Pre-recorded Crowd Noise through the Speakers at London Stadium, at www.dreamteamfc.com/c/archives/news-gossip/143103/west-ham-accused-playing-crowd-noise-speakers-attempt-create-atmosphere/, accessed 05-05-20.

Dykhoff, K. (2003), About the Perception of Sound, at www.dramatiskainstitutet.se/web/About_the_perception_of_sound.aspx, accessed 11-07-18.

Dziadul, C. (2016), BT Sport to Offer Dolby Atmos Sound, at www.broadbandtvnews.com/2016/12/09/bt-sport-offer-dolby-atmos-sound/, 09-12-16, accessed 26-07-17.

Edgar, R. (2010), *The Language of Film*, London: Fairchild Books.

Ekman, I. (2014), A Cognitive Approach to the Emotional Function of Game Sound, in *The Oxford Handbook of Interactive Audio*, Collins, et al. (eds), pp.196–212, New York: Oxford University Press.

Ekman, P., Sorenson, E. R. & Friesen, W. V. (1969), Pan-Cultural Elements in Facial Displays of Emotions, *Science*, 164, pp. 86–88.

Ekman, P. (1999), Basic Emotions, in *Handbook of Cognition and Emotion*, Dalgleish, T. and Power, M. (eds), pp.45–60, New York: Wiley & Sons.

Fehr, B. & Russell, J A. (1984), Concept of Emotion Viewed from a Prototype Perspective, *Journal of Experimental Psychology: General*, 113(3), pp. 464–486.

Feldman Barrett, L. (2017), *How Emotions Are Made: The Secret Life of the Brain*, London: Macmillan.

Fleming, M. (2012), Oscars Q&A: Steven Spielberg on War Horse's Four-Legged Actors, 3D and Lessons Learned, *Deadline.com*, at https://deadline.com/2012/02/oscars-qa-s teven-spielberg-on-war-horses-four-legged-actors-3d-and-lessons-learned-223038/, accessed 03-07-20.

Freebets (2016), Embarrassing: West Ham Accused of Playing Pre-recorded Crowd Noise at Games to Improve Atmosphere at London Stadium, at www.freebets.co.uk/opinion/ embarrassing-west-ham-accused-playing-pre-recorded-crowd-noise-games-improve-a tmosphere-london-stadium/, accessed 05-05-20.

Frick, R. W. (1985), Communicating Emotion: The Role of Prosodic Features, *Psychological Bulletin*, 97(3), pp. 412–429.

Gallese, V., Fadiga, L., Fogassi, L. & Rizzolatti, G. (1996), Action Recognition in the Premotor Cortex, *Brain*, 119, pp. 593–609.

Gavinson, T. (2017), Keep on Fighting: An Interview with Malala Yousafzai, *Rookie*, at www.rookiemag.com/2017/10/keep-on-fighting-an-interview-with-malala-yousafzai/, accessed 20-07-2020.

Gerrig, R. J. (1996), The Resiliency of Suspense, in Vorderer, et al. (eds), *Suspense: Conceptualizations, Theoretical Analyses and Empirical Explorations*, Mahwah, NJ: Lawrence Erlbaum.

Gilbey, R. (2015), Robert Altman: The Genius Who 'Reinvented the Language of Cinema', *The Guardian*, at www.theguardian.com/film/2015/mar/19/robert-altman-genius-who-reinvented-language-of-cinema, accessed 31-07-18.

Gould, R. (2016), *Auditory Icons*, at http://designingsound.org/2016/11/auditory-icons/, accessed 24-07-17.

Gonzalez, R. (2020), Coronavirus: Manchester City's Guardiola Would Rather Have Games Called Off than Played Behind Closed Doors, at www.cbssports.com/soccer/ news/coronavirus-manchester-citys-guardiola-would-rather-have-games-called-off-t han-played-behind-closed-doors/, accessed 05-05-20.

Graham, S. A., San Juan, V. & Khu, M. (2016), Words Are Not Enough: How Preschoolers' Integration of Perspective and Emotion Informs Their Referential Understanding, *Journal of Child Language*, Cambridge: Cambridge University Press, at www.cambri dge.org/core/services/aop-cambridge-core/content/view/182E0342DA81D19A2A9F8 B29DB024644/S0305000916000519a.pdf/words_are_not_enough_how:preschoolers_ integration_of_perspective_and_emotion_informs_their_referential_understanding. pdf, accessed 15-04-17.

Grosz, C. (2011), Inside Aaron Sorkin's Writing Process, *Hollywood Reporter*, 08-01-2011 at www.hollywoodreporter.com/news/aaron-sorkins-writing-process-69586, accessed 19-05-17.

Hickman, M. (2011), 6 Things You Probably Didn't Know about '*It's a Wonderful Life*', at www.mnn.com/lifestyle/arts-culture/stories/6-things-you-probably-didnt-know-about-its-a-wonderful-life, accessed 04-05-20.

Higson, A. D. (1994), A Diversity of Film Practices: Renewing British Cinema in the 1970s, in *Catastrophe Culture? The Challenge of the Arts in the 1970s*, Gilbert, B. M. (ed.), pp. 216–239, Abingdon: Routledge.

Holland, N. (2009), *Literature and the Brain*, Gainesville: PsyArt Foundation.

Holman, T. (2002), *Sound for Film and Television*, Boston, MA: Focal Press.

Horn, J. (2010), Some Viewers Need a Hand after the Forearm Amputation in '127 Hours', *Los Angeles Times*, 31-10-2010 at http://articles.latimes.com/2010/oct/31/entertainm ent/la-et-arm-movie-20101031, accessed 14-06-16.

Hunter, P. & Schellenberg, E. (2010), Music and Emotion, in *Music Perception*, Riess Jones, M., Fay, R. & Popper, A. (eds), pp. 129–164, New York: Springer.

Iacoboni, M. (2008), Mesial Frontal Cortex and Super Mirror Neurons, *Behavioral and Brain Sciences*, 31(1), pp. 30–30.

IMDb (Internet Movie Database) for *Walter Murch*, at www.imdb.com/name/nm0004555/#SoundDepartment, accessed 11-09-11.

Johanson, M. (2017), Finding Fatimah Movie Review: Dating while Muslim, at www.flickf ilosopher.com/2017/04/finding-fatimah-movie-review-dating-muslim.html, accessed 19-05-20.

Juslin, P. & Sloboda, J. (eds) (2010), *Handbook of Music and Emotion: Theory, Research, Applications*, Oxford: Oxford University Press.

Kahn, C. (2010), Erin Brockovich II? Activist Returns to Aid Town, at www.npr.org/2010/12/13/131967600/erin-brockovich-ii-activist-returns-to-aid-town?t=1588569705850, accessed 04-05-20.

Kayani, A. (2017), CloseShave TV Interview with *Finding Fatimah* Writer and Director, Oz Arshad, at www.youtube.com/watch?v=Kv1NGgI-Ev4, accessed 19-05-20.

Keh, A. (2020), We Hope Your Cheers for this Article Are for Real, at www.nytimes.com/2020/06/16/sports/coronavirus-stadium-fans-crowd-noise.html, accessed 27-09-20.

Kellaway, R. (2014), Eyes of the World on Glasgow as Scotland Hosts Commonwealth Games Opening Ceremony, at www.express.co.uk/news/uk/491099/Commonwealth-Games-2014-Opening-ceremony-Queen-Rod-Stewart-Susan-Boyle-TV-audience-bato n, 24/07/14, accessed 07-08-14.

Kenny, T. (2004), Gary Rydstrom, at http://groep6crm2b.wikidot.com/gary-rydstrom, February 1st, 2004, accessed 03-06-17.

Kermode and Mayo's Film Review (2014), BBC Radio 5 on-line 04-07-14, at www.bbc.co.uk/programmes/p0225p32, accessed 22-10-14.

Kermode and Mayo's Film Review (2017), BBC Radio 5 on-line 13-05-17, at www.bbc.co.uk/programmes/b08pgbr2, accessed 19-05-17.

Kermode, M. (2014), *Here and Now* Review, *The Observer Newspaper* on-line 06-07-14, at www.theguardian.com/film/2014/jul/06/here-and-now-review-lisle-turner-laure n-johns, accessed 22-10-14.

Keysers, C., Kohler, E., Umiltà, M. A., Nanetti, L., Fogassi, L. & Gallese, V. (2003), Audiovisual Mirror Neurons and Action Recognition, *Experimental Brain Research*, 153, pp. 628–636.

Knautz, K. (2012), Emotion Felt and Depicted: Consequences for Multimedia Retrieval, in *Indexing and Retrieval of Non-Text Information*, Rasmussen Neal, D. (ed.), pp. 343–375, Berlin: Walter de Gruyter.

Kornfield, J. (2014), *A Lamp in the Darkness: Illuminating the Path Through Difficult Times*, Boulder, CO: Sounds True.

Langridge, M. (2017), Football in 4K and Dolby Atmos, Is There Anything Better?, at www.pocket-lint.com/news/140141-football-in-4k-and-dolby-atmos-is-there-anything-better-our-verdict, 02-02-17, accessed 26-07-17.

Levy, E. (2009), New York, I Love You: Shekhar Kapur in the Upper East Side, *Cinema 24/7*, at https://emanuellevy.com/review/new-york-i-love-you-shekhar-kapur-in-the-upper-east-side-9/, accessed 10-06-20.

Loewinger, L. (1998), A Sound Idea: The Rationale Behind the Position of "Sound Designer" and Why it Never Took Hold, *The Independent* (the magazine of the Foundation for Independent Film and Video), October issue, New York: Foundation for Independent Video and Film, at www.larryaudio.com/aSouIde.html, accessed 12-05-17.

Lumet, S. (1995), *Making Movies*, London: Bloomsbury Publishing.

Marston, W. M. (1928), *Emotions of Normal People*, Abingdon: Routledge, Trench & Trubner.

McCann, G. (2007), *The New Introduction to Adorno, T. & Eisler, H.*, Composing for the Films, London: Continuum.

Michael (1996), Theory of Electromechanical Switching, at www.seg.co.uk/telecomm/automat1.htm, accessed 10/04/2011.

Mottram, J. (2014), *Here and Now* review in *Total Film Magazine*, issue 222, p. 53.

Murch, W. (2005), Dense Clarity, Clear Density, at www.transom.org, accessed 05-01-16.

Murch, W. (2020), Quoting from private correspondence to the author.

Nelson, R. (2013), *Practice as Research in the Arts: Principles, Protocols, Pedagogies, Resistances*, Basingstoke: Palgrave Macmillan.

Neville, S. (2011), The Craftsman, in *Down These Green Streets*, Burke, D. (ed.), pp. 235–248 Dublin: Liberties Press.

Ondaatje, M. (2004), *The Conversations: Walter Murch and the Art of Editing Film*, London: Bloomsbury.

Pauletto, S. (2012), The Sound Design of Cinematic Voices, *The New Soundtrack*, 2(2), pp. 127–142, Edinburgh: Edinburgh University Press.

Pereira, C. (2000), Dimensions of Emotional Meaning in Speech, *ISCA Workshop on Speech & Emotion*, Northern Ireland, pp. 25–28.

Plantinga, C. (2006), Disgusted at the Movies, *Film Studies*, Cinquegrani, M., Frey, M. & Kamm, F. et.al. (eds), pp. 81–92; Manchester: Manchester University Press.

Plunkett, J. (2014), BBC Olympics Coverage Watched by 90% of UK Population, at www.theguardian.com/media/2012/aug/13/bbc-olympics-coverage-tv-ratings,   13/08/12, accessed 07-08-14.

Plutchik, R. (2001), The Nature of Emotions, *American Scientist*, 89(4), p. 344, doi: 10.1511/2001.4.344.

Rizzolatti, G., Fadiga, L., Fogassi, L. & Gallese, V. (1996), Premotor Cortex and the Recognition of Motor Action, *Cognitive Brain Research*, 3, pp. 131–41.

Robinson, J. (2005), *Deeper than Reason: Emotion and Its Role in Literature, Music, and Art*, Oxford: Oxford University Press.

Ramachandran, V. (2009), The Neurons that Shaped Civilization, *Trans EDUC India 2009*, at www.ted.com/talks/vs_ramachandran_the_neurons_that_shaped_civilization?language=en#t-442466, accessed 05-01-16.

Renoir, J. (1974), *My Life and My Films*, Boston, MA: Da Capo Press.

Rozett, M. (1998), Interview with Richard Portman, at http://filmsound.org/richardportman/richardportman.htm, accessed 09-05-20.

Russell, J. A. (1980), A Circumplex Model of Affect, *Journal of Personality and Social Psychology*, 39, pp. 1161–1178.

Sacharin, V., Schlegel, K. & Scherer, K. R. (2012), *Geneva Emotion Wheel Rating Study* (Report). Geneva, Switzerland: University of Geneva, Swiss Centre for Affective Sciences.

Sandbrook, D. (2011), *State of Emergency: The Way We Were: Britain, 1970–1974*, London: Penguin.

Scherer, K. R. (2005), What Are Emotions? And How Can They Be Measured?, *Social Science Information*, 44(4), pp. 695–729.

Scherer, K. R., Shuman, V., Fontaine, J. R. J. & Soriano, C. (2013), The GRID Meets the Wheel: Assessing Emotional Feeling via Self-report, in *Components of Emotional Meaning: A Sourcebook* Johnny R. J. Fontaine, Klaus R. Scherer, & C. Soriano (eds), pp. 281–298. Oxford: Oxford University Press.

Schrimshaw, W. (2013), Non-cochlear Sound: On Affect and Exteriority, in *Sound, Music, Affect: Theorizing Sonic Experience*, Thompson, M. & Biddle, I. (eds), p. 27, New York: Bloomsbury Academic.

Sergi, G. (2004), *The Dolby Era: Film Sound in Contemporary Hollywood*, Manchester: Manchester University Press.

Serres, M. (2012), *Biogée*, Minneapolis, MN: Univocal Publishing.

Shakespeare, W. (1600), *A Midsummer Night's Dream*, Act 2, Scene 1, p. 8 at http://nfs.sparknotes.com/msnd/page_46.html, accessed 28-07-18.

Shaver, P., Schwartz, J., Kirson, D. & O'Connor, C. (1987), Emotion Knowledge: Further Exploration of a Prototype Approach, *Journal of Personality and Social Psychology*, 52, pp. 1061–1086.

Shaviro, S. (2016), Affect vs Emotion, in *The Cine-Files*, issue 10, at www.thecine-files.com/shaviro2016/, accessed 11-07-18.

Shergold, A. (2020), Sky Will Draw on 1,300 Different Chants, Songs and Crowd Noise Samples from EA Sports' FIFA Archive to Enhance Premier League Coverage as Season Resumes Behind Closed Doors (and Don't Worry, Swearing HAS Been Filtered Out!), at www.dailymail.co.uk/sport/sportsnews/article-8425897/Sky-Sports-draw-sound-archive-FIFA-games-enhance-Premier-League-coverage.html, accessed 27-09-20.

Shouse, E. (2005), Feeling, Emotion, Affect, *M/C Journal*, 8(6), at http://journal.media-culture.org.au/0512/03-shouse.php, accessed 11-07-18.

Sider, L. (2003), If You Wish to See, Listen, *Journal of Media Practice*, 4(1), pp. 5–15, Bristol: Intellect.

Siegel, S. B. (2004), *The Encyclopaedia of Hollywood*, p. 256, New York: Checkmark Books.

Silver, J. (2005), Exploitation is More Widespread than Ever, at www.theguardian.com/media/2005/apr/11/broadcasting.mondaymediasection, accessed 15-03-20.

Smith, H. & Dean, R. T. (eds) (2009), *Practice-Led Research, Research-Led Practice*, Edinburgh: Edinburgh University Press.

Sonnenschein, D. (2001), *Sound Design: The Expressive Power of Music, Voice and Sound Effects in Cinema*, Studio City: Michael Wiese Productions.

Stam, R. & Miller, T. (eds) (2000), *Film Theory: An Introduction*, Oxford: Blackwell Publishing Ltd.

Storr, G. (2017), Film Review: Finding Fatimah, *The National Student*, at www.thenationalstudent.com/Film/2017-04-18/film_review:finding_fatimah.html, accessed 19-05-20.

Sullivan, B. (2012), Making the Olympics Sound Right, From a 'Swoosh' to a 'Splash', at www.npr.org/sections/thetorch/2012/07/28/157442046/making-the-olympics-sound-right-from-a-swoosh-to-a-splash, accessed 26-07-17.

Swaminathan, S. & Schellenberg, E. G. (2015), Current Emotion Research in Music Psychology, *Journal of the International Society for Emotional Research*, 7(2), at http://journals.sagepub.com/doi/abs/10.1177/1754073914558282?journalCode=emra, accessed 13-04-17.

Tarkovsky, A. (1987), *Sculpting in Time: Reflections on the Cinema* (Trans. Hunter-Blair, K.) at https://monoskop.org/images/d/dd/Tarkovsky_Andrey_Sculpting_in_Time_Reflections_on_the_Cinema.pdf, accessed 28-07-18.

Terranova, A. (2014), Italian Voice Actors Show Their Faces, *The New Yorker Magazine*, at www.newyorker.com/culture/photo-booth/italian-voice-actors-show-their-faces, accessed 30-05-20.

Théberge, P. (2008), *Lowering the Boom: Critical Studies in Film Sound*, Beck, J. & Grajeda, T. (eds), Chicago, IL: University of Illinois.

The Swedish Film Database (1963), Winter Light, at www.svenskfilmdatabas.se/en/Item/?type=film&itemid=4663, accessed 30-07-18.

Thom, R. (1998), More Confessions of a Sound Designer (A Sound Fails in The Forest Where Nobody Hears It), at http://filmsound.org/randythom/confess2.html, accessed 07-05-20.

Thom, R. (2011), Screenwriting for Sound, *The New Soundtrack 1.2*, Edinburgh: Edinburgh University Press.

Tookey, C. (2010), Slumdog Was Feelgood … This Blood-Stained Film is Feelbad, at www. dailymail.co.uk/tvshowbiz/article-1324706/Danny-Boyles-127-Hours-premieres-London-Film-Festival.html, accessed 12-05-17.

Treasure, J. (2007), *Sound Business*, Cirencester: Management Books.

Van Bezooijen, R. (1984), *Characteristics and Recognizability of Vocal Expressions of Emotion*, Dordrecht (Holland): Foris Publications.

Van Luling, T. (2014), 5 Things You Didn't Know about 'Jurassic Park' at www.huffin gtonpost.com/2014/12/18/jurassic-park-trivia_n_6329640.html, accessed 12-05-17.

Ward, M. (2015), Art in Noise, in *Embodied Cognition and Cinema*, Coëgnarts, M. & Kravanja, P. (eds), pp. 155 – 186, Leuven: Leuven University Press.

Weninger, F., Eyben, F., Schuller, B. W., Mortillaro, M. & Scherer, K. R. (2013), On the Acoustics of Emotion in Audio: What Speech, Music, and Sound Have in Common, *Frontiers in Psychology | Emotion Science*, 4(292), at www.ncbi.nlm.nih.gov/pmc/artic les/PMC3664314/, accessed 05-01-16.

Whittington, W. (2007), *Sound Design & Science Fiction*, Austin, TX: University of Texas.

Wigmore, T. (2020), Jofra Archer Calls for Artificial Crowd Noise at England Matches Behind Closed Doors, at www.telegraph.co.uk/cricket/2020/05/13/jofra-archer-calls-ar tificial-crowd-noise-england-matches-behind, accessed 14-05-20.

Willems, C. (2010), But What Makes It Doctoral? Taking on the Traditionalists: Interdisciplinary, Practice-Led Doctoral Research in the Creative Industries – A Case Study in Academic Politics, Research, Rigour and Relevance, *International Journal of Interdisciplinary Social Sciences*, 5(7), pp.331–345.

Wright, B. (2008), Aspect Ratio, at https://aspectratio.wordpress.com/2008/10/22/from-here-on-in-absolute-silence/, accessed 31/07/18.

Yahoo (2011), 127 Hours: A Film to Make You Faint – Yahoo!7 Movies, at http://au. movies.yahoo.com/on-show/article/-/8810220/127-hours-a-film-to-make-you-faint/, accessed 13-08-13.

Zanotti, S. (2015), Analysing Redubs: Motives, Agents and Audience Response, in *Audiovisual Translation in a Global Context: Mapping an Ever-changing Landscape*, Piñero, R. & Cintas, J. (eds), pp. 110–139, Basingstoke: Palgrave Macmillan.

Zentner, M. & Eerola, T. (2010), *Handbook of Music and Emotion: Theory, Research, Applications*. Juslin, P. & Sloboda, J. (eds), Oxford: Oxford University Press.

Zuckerman, D. (2009), Review: New York, I Love You, *Film Comment*, at www.filmco mment.com/article/new-york-i-love-you-review/, accessed 10-06-20.

Film and television referred to in this book

28 Days Later (2002), [Film]. D. Boyle. dir. UK: DNA Films.
2001: A Space Odyssey (1968), [Film]. S. Kubrick. dir. USA: Metro- Goldwyn-Mayer.
3 Women (1977), [Film]. R. Altman. dir. USA: Lion's Gate Films.
A Christmas Carol (2019), [TV series]. N. Murphy. dir. UK: BBC.
A Clockwork Orange (1971), [Film]. S. Kubrick. dir. USA: Warner Bros.
A Few Good Men (1992), [Film]. R. Reiner. dir. USA: Columbia Pictures.

Afterglow (1997), [Film]. A. Rudolph. dir. USA: Elysian Dreams.
Almost Famous (2000), [Film]. C. Crowe. dir. USA: Dreamworks/Columbia Pictures.
American Graffiti (1973), [Film]. G. Lucas. dir. USA: Universal Pictures.
Antiques Roadshow (1979), [TV series]. D. Hess et al. dir. UK: BBC.
Apocalypse Now (1979), [Film]. F. F. Coppola. dir. [DVD Redux edition, 2002] USA: Zoetrope Studios.
A Serious Man (2009), [Film]. J. and E. Coen. dirs. [DVD 2010] USA: Focus Features.
Away from Her (2006), [Film]. S. Polley. dir. Canada: Foundry Films.
Barry Lyndon (1975), [Film]. S. Kubrick. dir. UK: Peregrine.
Barton Fink (1991), [Film]. J. and E. Coen. dirs. USA: Circle Films.
Batman Returns (1992), [Film]. T. Burton. dir. USA: Warner Bros.
Bend It Like Beckham (2002), [Film]. G. Chadha. dir. USA: Kintop Pictures.
Buffy the Vampire Slayer (1997–2003), [TV series]. J. Whedon, J. Contner, D. Solomon et al. dirs. USA: Mutant Enemy.
Butt Ugly Martians (2001), [TV series]. J. Prikryl et al. dir. USA: DCDC.
Beyond the Horizon (2000), [TV series]. M. Clements. dir. UK: Discovery.
Bicentennial Man (1999), [Film]. C. Columbus. dir. USA: 1492 Pictures.
Bridget Jones's Diary (2001), [Film]. S. Maguire. dir. USA: Miramax.
Bridget Jones: The Edge of Reason (2004), [Film]. B. Kidron. dir. USA: Miramax.
Charlie Wilson's War (2007), [Film]. M. Nichols. dir. USA: Universal Pictures.
Citizen Kane (1941), [Film]. O. Welles. dir. USA: RKO Radio Pictures.
Côte d'Azur (2005), [Film]. O. Ducastel & J. Martineau. dirs. France: Agat Films & Cie.
Dad's Army (1968–1977), [TV series]. H. Snoad et al. dir. UK: BBC.
Darling (1965), [Film]. J. Schlesinger. dir. UK: Joseph Janni Production.
Dogville (2003), [Film]. [DVD 2010]. L. von Trier. dir. Denmark: Zentropa Entertainments.
Don Juan (1926), [Film]. A. Crosland. dir. USA: Warner Bros.
Dr. Strangelove (1964), [Film]. S. Kubrick. dir. USA: Columbia Pictures.
Dr. Zhivago (1965), [Film]. D. Lean. dir. USA: Metro-Goldwyn-Mayer.
Dunkirk (2017), [Film]. C. Nolan. dir. USA: Warner Bros.
East Is East (1999), [Film]. D. O'Donnell. Dir. UK: FilmFour.
Enter the Dragon (1973), [Film]. R. Clouse. dir. USA: Warner Bros.
Eraserhead (1977), [Film]. D. Lynch. dir. USA: American Film Institute.
Erin Brockovich (2000), [Film]. S. Soderbergh. dir. USA: Universal Pictures.
Fantasia (1940), [Film]. J. Algar, S. Armstrong. dirs. USA: Walt Disney Productions.
Fargo (1996), [Film]. [DVD 2007]. J. and E. Coen. dirs. USA: PolyGram Filmed Entertainment.
Fifth Gear (2002), [TV series]. M. McQueen et al. dir. UK: North One Television.
Finding Fatimah (2017), [Film]. O. Arshad. dir. UK: British Muslim TV.
Garden State (2004), [Film]. Z. Braff. dir. USA: Camelot Pictures.
Gosford Park (2001), [Film]. R. Altman. dir. USA: USA Films.
Half A Sixpence (1967), [Film]. G. Sidney. dir. UK: Ameran Films.
Hello, Dolly! (1969), [Film]. G. Kelly. dir. USA: Twentieth Century Fox.
Here and Now (2014), [Film]. L. Turner. dir. UK: Wrapt Films.
Holiday on the Buses (1973) [Film]. B. Izzard. dir. UK: Anglo-EMI.
Home from Home (2002–2003) [TV series]. S. Brooks et al. dirs. UK: Maverick Television.
Horizon (1964), [TV series]. R. A. Sisson et al. dir. UK: BBC.
How to Train Your Dragon (2010), D. DeBlois & C. Sanders. dirs. USA: DreamWorks Animation.

Inception (2010), [Film]. C. Nolan. dir. USA: Warner Bros.

It's a Wonderful Life (1946), [Film]. [DVD, 2007]. F. Capra. dir. USA: RKO Radio Pictures.

Jack - Safe@Last (2010), [Online video] J. Stinton. dir. UK: ST16.

King Kong (1933), [Film] M. C. Cooper, E. B. Schoedsack. dirs. USA: RKO Radio Pictures.

King of the Hill (1997–2010), [TV series]. W. Archer et al. dir. USA: Deedle-Dee Productions.

Kingsman: The Golden Circle (2017), [Film]. M. Vaughn. dir. USA: Twentieth Century Fox.

Kung Fu (1972–1975) [TV series]. J. Thorpe et al. dir. USA: Warner Bros.

La Notte (1961), [Film]. M. Antonioni. dir. Italy: Nepi Film.

Last Shop Standing (2012), [Film]. P. Piper. dir. UK: Blue Hippo Media.

Lawrence of Arabia (1962), [Film]. D. Lean. dir. UK: Horizon Pictures.

Life of Brian (1979), [Film]. T. Jones. dir. UK: Hand Made Films.

Lincoln (2012), [Film]. S. Spielberg. dir. USA: DreamWorks.

Lord of The Rings (2001), [Film]. P. Jackson. dir. USA: New Line Cinema.

Love Thy Neighbour (1972–1976), [TV series]. S. Allen et al. dir. UK: Thames Television.

Love Thy Neighbour (1973), [Film]. J. Robins. dir. UK: Anglo-EMI.

McCabe & Mrs. Miller (1971), [Film]. R. Altman. dir. USA: David Foster Productions.

Memphis Belle (1990), [Film]. M. Caton-Jones. dir. UK: Enigma Productions.

Minority Report (2002), [Film]. [DVD, 2003]. S. Spielberg. dir. USA: Twentieth Century Fox.

Mission: Impossible (1996), [Film]. B. De Palma. dir. USA: Paramount Pictures.

Money Kills (2012), [Film]. L. Murphy. dir. UK: Sheringham Studios.

Mountain Biking: The Untold British Story (2016), [Film]. M. B. Clifford. dir. UK: Blue Hippo Media.

Mutiny on the Buses (1972), [Film]. H. Booth. dir. UK: Anglo-EMI.

Nashville (1975), [Film]. R. Altman. dir. USA: Paramount Pictures.

New York, I Love You (2008), [Film]. A. Minghella (with S. Kapur). dirs. USA: Vivendi Entertainment.

No Country for Old Men (2007), [Film]. [DVD, 2008]. J. and E. Coen. dirs. USA: Paramount Vantage.

Notes from a Small Island (1999), [TV series]. R. Lightbody. dir. UK: Carlton Television.

Oceans 11 (2001), [Film]. S. Soderbergh. dir. USA: Warner Bros.

On Dangerous Ground (1996), [Film]. L. Gordon-Clark. dir. USA: Carousel.

On the Beach (1959), [Film]. S. Kramer. dir. USA: Stanley Kramer Productions.

On the Buses (1969–1973), [TV series]. S. Allen et al. dir. UK: London Weekend Television.

Persona (1966), [Film]. I. Bergman. dir. Sweden: Svensk Filmindustri.

Pirates of the Caribbean: Dead Men Tell No Tales (2017), [Film]. Dir. E. Sandberg. dir. USA: Walt Disney Pictures.

Please Sir! (1968–1972), [TV series]. M. Stuart et al. dir. UK: London Weekend Television.

Please Sir! (1971), [Film]. M. Stuart. dir. UK: Rank Organisation.

Porridge (1974–1977), [TV series]. S. Lotterby. dir. UK: BBC.

Ready Player One (2018), [Film]. S. Spielberg. dir. USA: Warner Bros.

Rear Window (1954), [Film]. A. Hitchcock. dir. USA: Alfred J. Hitchcock Productions.

Rififi (1955), [Film]. J. Dassin. dir. France: Pathé Consortium Cinéma.

Rising Damp (1974–1978), [TV series]. R. Baxter et al. dir. UK: Yorkshire Television.

Saving Private Ryan (1998), [Film]. S. Spielberg. dir. USA: DreamWorks.

Scott and Sid (2015), [Film]. S. Elliott. dir. UK: Dreamchasers Films.

Seven Samurai (1954), [Film]. A. Kurosawa. dir. Japan: Toho Company.

Slumdog Millionaire (2008), [Film]. D. Boyle. dir. UK: Celador Films.

Solaris (2002), [Film]. S. Soderbergh. dir. USA: Twentieth Century Fox.

Solyaris (1972), [Film]. A. Tarkovsky. dir. USSR: Mosfilm.

Spitting Image (1984–1996), [TV series]. P. Harris et al. dir. UK:Central Television.

Star Trek (1966–1969), [Original TV series]. M. Daniels et al. dir. USA: Desilu Productions.

Star Wars: Episode IV – A New Hope (1977), [Film]. G. Lucas. dir. USA: Lucasfilm.

Star Wars: Episode V – The Empire Strikes Back (1980), [Film]. G. Lucas. dir. USA: Lucasfilm.

Star Wars: Episode VIII – The Last Jedi (2017), [Film]. R. Johnson. dir. USA: Lucasfilm.

Steptoe and Son (1962–1974), [TV series]. D. Wood et al. dir. UK: BBC.

Steptoe and Son (1972), [Film]. C. Owen. dir. UK: Associated London Films.

Superman (1978), [Film]. R. Donner. dir. UK: Dovemead Films.

The Aristocats (1970), [Film]. W. Reitherman. dir. USA: Walt Disney Productions.

The Aviator (2004), [Film]. M. Scorsese. dir. USA: Forward Pass.

The Big Lebowski (1998), [Film]. J. and E. Coen. dirs. USA: Polygram Filmed Entertainment.

The Conversation (1974), [Film]. F. F. Coppola. dir. USA: The Directors Company.

The Craftsman (2012), [Film]. L. Murphy. dir. UK: Sheringham Studios.

The Elephant Man (1980), [Film]. D. Lynch. dir. USA: Brooksfilms.

The English Patient (1996), [Film]. [DVD, 2005]. A. Minghella. dir. USA: Miramax.

The French Lieutenant's Woman (1981), [Film]. K. Reisz. dir. UK: Juniper Films.

The Gadget Show (2004), [TV series]. D. Leighton et al. dir. UK: North One Television.

The Godfather Part II (1974), [Film]. F. F. Coppola. dir. USA: Paramount Pictures.

The Godfather Part III (1990), [Film]. F. F. Coppola. dir. USA: Paramount Pictures.

The Incredibles (2004), [Film]. B. Bird. dir. USA: Pixar Animation Studios.

The Jazz Singer (1927), [Film]. A. Crosland. dir. USA: Warner Bros.

The Levelling (2016), [Film]. H. Dickson Leach. dir. UK: Oldgarth Media.

The Likely Lads (1964–1966), [TV series]. D. Clement. dir. UK: BBC.

The Likely Lads (1976), [Film]. M. Tuchner. dir. UK: Anglo-EMI Productions.

The Mirror (1975), [Film]. A. Tarkovsky. dir. USSR: Mosfilm.

The Polar Express (2004), [Film]. R. Zemeckis. dir. USA: Castle Rock Entertainment.

The Rainmaker (1997), [Film]. F. F. Coppola. dir. USA: Constellation Entertainment.

The Rain People (1969), [Film]. F. F. Coppola. dir. USA: American Zoetrope.

The Right Stuff (1983), [Film]. P. Kaufman. dir. USA: Ladd Company.

The Robe (1953), [Film]. H. Koster. dir. USA: Twentieth Century Fox.

The Seventh Seal (1957), [Film]. I. Bergman. dir. Sweden: Svensk Filmindustri

The Silence. 1963. [Film]. I. Bergman. dir. Sweden: Svensk Filmindustri.

The Shining (1980), [Film]. S. Kubrick. dir. USA: Warner Bros.

The Simpsons (1989), [TV series]. D. Silverman et al. dir. USA: Gracie Films.

The Social Network (2010), [Film]. D. Fincher. dir. USA: Columbia Pictures.

The Sopranos (1999–2007), [TV series]. T. Van Patten, J. Patterson, A. Coulter et al. dirs. USA: HBO.

The Thomas Crown Affair (1968), [Film]. N. Jewison. dir. USA: Mirisch Corporation.

The X Files (1993–2018), [TV series]. K. Manners, R. Bowman et al. dirs. USA: Ten Thirteen Productions.

Throne of Blood (1957), [Film]. A. Kurosawa. dir. Japan: Toho Company.

Through a Glass Darkly. 1961. [Film]. I. Bergman. dir. Sweden: Svensk Filmindustri.

THX 1138 (1971), [Film]. G. Lucas. dir. USA: American Zoetrope.
Tomorrow's World (1965–2003), [TV series] P. R. Smith et al. dir. UK: BBC.
Top Gear (1978–2002), [Original TV series] D. Wheeler, D. Jarvis et al. dirs. UK: BBC.
Top Gun (1986), [Film]. T. Scott. dir. USA: Paramount Pictures.
Tora! Tora! Tora! (1970), [Film]. R. Fleischer, K. Fukasaku, T. Masuda. dirs. USA: Twentieth Century Fox.
Touch of Evil (1958), [Film]. O. Welles. dir. USA: Universal Pictures.
True Grit (2010), [Film]. J. and E. Coen. dirs. USA: Paramount Pictures.
Up Pompeii! (1969–1970), [TV series]. D. Croft et al. dir. UK: BBC.
Up Pompeii (1971), [Film]. B. Kellett. dir. UK: Anglo-EMI.
Wall-E (2008), [Film]. A. Stanton. dir. USA: Pixar Animation Studios.
War Horse (2011), [Film]. S. Spielberg. dir. USA: DreamWorks.
Wayne's World (1992), [Film]. P. Spheeris. dir. USA: Paramount Pictures.
When Harry Met Sally (1989), [Film]. R. Reiner. dir. USA: Castle Rock Entertainment.
Where Eagles Dare (1968), [Film]. G. B. Hutton. dir. USA: Gershwin-Kastner Productions.
Winter Light (1963), [Film]. [DVD, 2001]. I. Bergman. dir. Sweden: Svensk Filmindustri.

Music referred to in this book

Pet Sounds. 1966. [Album]. The Beach Boys. USA: Capitol Records.
Promises and Lies. 1993. [Album]. UB40. UK: Virgin.
Sgt. Pepper's Lonely Hearts Club Band. 1967. [Album]. The Beatles. UK: Parlophone.

Index

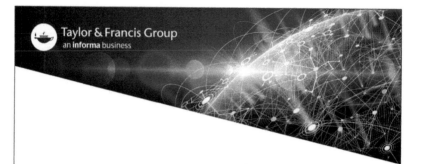